Foundations of Full Text Electronic Information Delivery Systems

IMPLICATIONS FOR INFORMATION PROFESSIONALS

Harry M. Kibirige

Neal-Schuman Publishers, Inc.
New York *London*

Published by Neal-Schuman Publishers, Inc.
100 Varick Street
New York, NY 10013

Printed and bound in the United States of America

Library of Congress Cataloging-in-Publication Data

Kibirige, Harry M.
Foundations of full text electronic information delivery systems / by Harry M. Kibirige
p. cm.
Includes bibliographical references and index.
ISBN 1-55570-208-2
1. Facsmilile transmission. I. Title.
TK6710.K53 1996
004.6'9—dc20 96-28444

Dedication

To: my wife Christine, and my children: Diane, Harry Jason Jr., Jane, and Sarah

Contents

Acknowledgments

My special thanks go to trustees of the City University of New York Research Foundation for the grant that enabled me to conduct the basic field research needed for this book. I am indebted to my research assistants Jennie Hazan and Varghese Mathai for providing such invaluable help in data collection and input. To Ben Chitty and Jean Petrisek, I extend my sincere thanks for thoroughly reading the manuscript and providing suggestions for improvement. Thanks to Stephanie Goldson, for such a wonderful set of illustrations that make the book attractive and informative. I thank the Queens College GLIS faculty for their understanding and the Inter-Library Loan Department for getting me research materials on demand. It made my life immeasurably easier. Many thanks go to Walter Spilsbury for his warm friendship and help in searching and obtaining a habitable environment for my family. Finally, heartfelt thanks to my wife and children for their patience, encouragement, love, and understanding while I worked on an apparently endless project.

Figures

Preface

As the twentieth century dovetails into the twenty first century, information has become an increasingly valuable commodity. Access to full text information, containing text, images, and in some cases, sound, is becoming vital to decision-making for organizations as well as individuals. With *Foundations of Full Text Information Delivery Systems* I aim to examine the information phenomena as it affects human society.

The stimulus to write this book came after I conducted a nationwide study on electronic information delivery systems, funded by the City University of New York Research Foundation. My findings reveal a very keen interest in full text information access by information professionals. Programs at conferences also reflect the same sentiments.

While information professionals are concerned with full text information delivery systems, they also are concerned with the enabling technology underlying them, namely telecommunications networks. They want to know the *means* behind remote information access. This is why information delivery systems and the telecommunications infrastructure form the first of two basic themes. The human factors associated with information access comprises the third but essential theme both of *Foundations of Full Text Electronic Information Delivery Systems* and of the role of full text delivery in libraries and other organizations.

With these themes in mind, I have written this book for both information professionals and information users. Educators in schools that emphasize information systems and science will find the book timely and useful as it highlights the centrality of full text information in organizations. Management Information Systems (MIS) programs in business schools will also find it useful as it incorporates information systems into the broad corporate culture (especially the role of information in the process). Other people interested in information infrastructure will find this book an essential companion as it explains how information from remote databanks gets to the individual's computer at home or work.

I used nontechnical language throughout to make this book readable for this broad audience. In the cases where technical terms were unavoidable, I defined them in the text when first introduced. The glossary of terms and acronyms at the end of the book is included for easy reference.

Chapter coverage starts with a general overview of the information marketplace in Chapter One. Trends, which I dub "cybertrends," are discussed here to set the stage for the rest of the book. Chapter One also emphasizes the supremacy of information in modern society.

Chapters Two and Three cover basic telecommunications technology and how systems and networks interconnect. Chapter Two is a tutorial on telecommunications written for the nontechnical reader, while Chapter Three discusses the rationale and crucial need for interconnectivity.

Chapter Four covers national and regional networks and how they facilitate information flow. Chapter Five also discusses networks, bringing them closer to the organizational or campus level where they are used as tools for achieving organizational objectives.

Chapter Six introduces intelligent gateway interfaces and covers methods of internetworking. This chapter reveals how gateways have facilitated seamless multiple system access using one terminal interface in a transparent mode.

U.S. national information policy is the focus of Chapter Seven, covering the elements of policy formulation and the major players in the telecommunications infrastructure. It gives an overview of the National Information Infrastructure (NII) in the context of full text information delivery.

Chapter Eight covers electronic information service providers, their tools of the trade (foundation) and the nature of their business. Appendix A supplements this chapter by providing names and addresses of these providers.

Full text delivery at the international level is covered in Chapter Nine. Problems and issues associated with national sovereignty, national boundaries, and the north-south controversy are treated here. The pros and cons of international transborder data flow (TBDF) are also covered.

Chapter Ten explores the human factors associated with information access. It stresses the need for equity in accessing information. It is assumed that most information systems are designed with the explicit or implicit purpose of improving the human condition.

Chapter Eleven summarizes the themes covered in *Foundations of Full Text Electronic Information Delivery Systems*. This chapter re-emphasizes the need for developing policies and designing systems for electronic full text information delivery for all. While skeptics may indeed disagree with the premise that such equity in information access can never be a reality since information technology has prospered only in capitalist societies.

While the reader of this book covers the technical foundations of information technology, I urge readers to perceive information as a means to an end rather than as an end unto itself. I believe that end to be the overall improvement of the quality of life.

1. The Information Market Place in a Cyberculture

Welcome to cyberculture! The revolutionary events of the last three decades are catapulting both the information user and the information provider into the twenty-first century practically unaware of broader implications underlying the information age. It is unnerving for both the professional and the layperson to witness, let alone experience this information environment. First, and foremost, a new terminology must be defined. Overused professional terms are not only clichés, but are also inadequate to describe the phenomenon that has come to be known as *cyberspace*. Alas, librarians, computer scientists, and other information professionals have to learn new terms that describe somewhat similar, but not identical, concepts. For instance, the technical usage of the term "folder" for Microsoft Windows and the Apple Computer system is similar to the colloquial usage for the manual filing cabinet system. However, the concept has a broader connotation in the computer environment.

This chapter sets the scene for the rest of the book by describing, explaining, and elucidating trends in what is often referred to as the information infrastructure. For lack of a better phrase and to persuade information specialists to stretch their collective imagination in order to navigate the information superhighways into the twenty first century, the term *cybertrends* will be adopted to reflect the dynamism and the state of flux of the forces of change. It is derived from the term cyberspace, which describes the modern world in which computers, telecommunications, and people coexist. As a corollary, the term *cybersources* refers to sources accessed to satisfy information needs as users navigate cyberspace. Cybertrends will profoundly affect full text information delivery systems which ultimately facilitate access to information on demand.

CYBERTRENDS

There are six major trends that will propel the information industry into the twenty first century. Not all of these motivating forces are new, but all are dynamic and volatile. Understanding their role in cyberspace is imperative in our attempt to visualize the nature of full text information delivery. The six major cybertrends are:

1. Miniaturization of the computing components and the concurrent enhancement of personal computing power.
2. Implementation of national and international telecommunications networks capable of handling gigabytes of data.
3. Research advances in electronic digital imaging.
4. Gradual transformation of traditional databases into cybersources.
5. Evolution of the cybersource user.
6. Full realization of Marshall McLuhan's global village prophesy through cyberculture within cyberspace.

These trends originated in the fourth quarter of the twentieth century and have attained incredible momentum. The development and utilization of full text information delivery systems is deeply intertwined with them. They are discussed in the following sections of this chapter in order to set the rest of the book in its true cyberculture perspective.

Diminutive but Powerful PCs in Cyberculture

From desktops and workstations to notebooks and personal digital assistants, that little device initially called the "personal computer," or simply PC, is neither personal nor merely a simple computer anymore. They are little in physique but are progressively attaining "giant" proportions with regard to computing power and storage. Any computer superstore sells several notebooks that can do what minicomputers and mainframes did a decade ago. The personal computer that used to focus on individual needs like balancing a checkbook, playing computer games, or producing personal memos, is now the main workhorse used by office workers and telecommuters to access work-related information sources. Since it can upload and download files from remote sites using sophisticated internal adapters, external gateways, and software, it is more of a transceiver—a device that receives and transmits digital signals within the dynamic information chain. The PC is making its indelible impact on cyberspace in six main areas.

1. The tremendous impact of PCs on the computer and information industry in the last two decades has shaken the ranks of minicomputer and mainframe manufacturing companies such as IBM, Digital Equipment Corporation, and Wang. Distributed processing, downsizing, and corporate mergers may to a certain extent be attributed to the momentous growth of the personal computer market within the last 20 years. As depicted in Figure 1.1, personal computer sales have soared since PCs were put on the market in the 1970s. Between 1981 and 1991, the compounded growth sales for PCs was over 20 percent. In contrast, sales of minicomputers and mainframes grew approximately 6 percent and 4 percent respectively. The PC has gradually become the computer of choice not only for individual users but also for corporations. When used in a local area network (LAN) environment, there are very few computer-driven tasks which the PC cannot do. With the introduction of the 32–bit EISA processor chips architecture, PCs are becoming formidable competitors for both minicomputers and mainframes. They process large text files better than the earlier generation of 16-bit, ISA-based chips. Coupled with gigabyte storage capacities and relatively large memories, PCs can now effectively process graphics and text. Medium sized full text processing is gradually migrating downward from minicomputers and mainframes to the personal computer workstation.

2. The PC has maintained its lead over dumb terminals as an easy-to-use access device for office and home use. Systems developers have appropriately responded to user demand by creating user friendly graphical user interfaces (GUIs). The Apple desktop concept pioneered the general public use of icon based access to user files which has been further developed and expanded by the Microsoft operating environments. Such systems appear less intimidating to first-time computer users and are preferred by many veteran computer users as well. By providing icons and mnemonic file-naming conventions, GUIs make computing environments congenial to most nontechnical users. The average full text user is nontechnical in contrast to users of scientific applications. Further development and widespread use of GUIs will make interfaces to full text systems more user friendly and thus more accessible to the general public.

Microcomputer Sales 1984–1994
Dollars in Millions
0
5000
10000
15000
20000
25000
1984
1985
1986
1987
1988
1989
1990
1991
1992
1993
1994

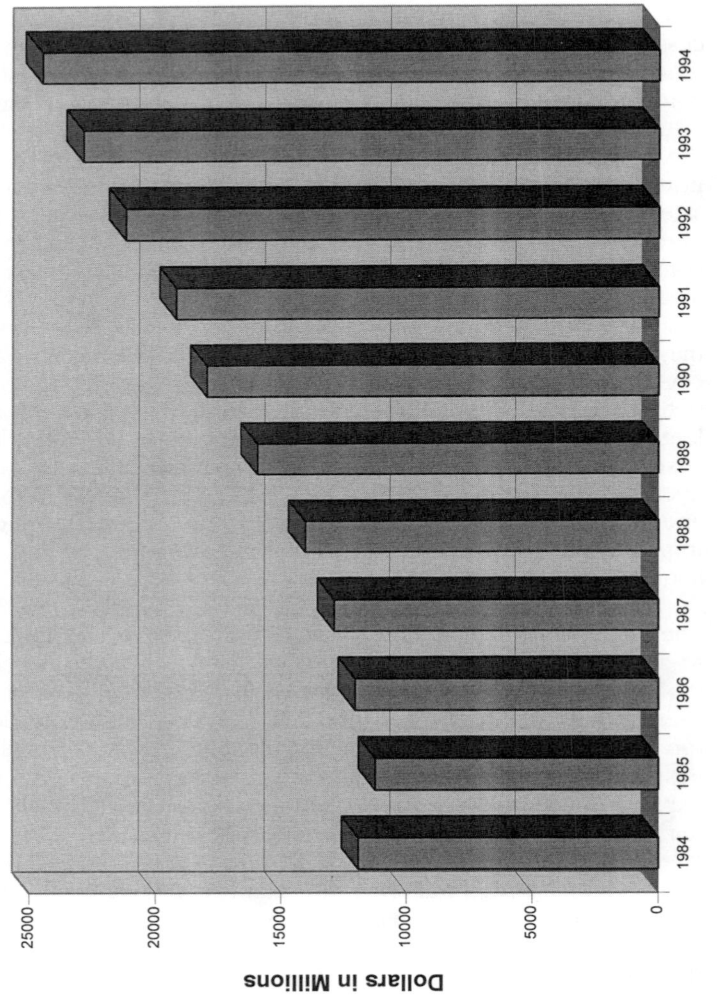

Figure 1.1

3. Spreading computer cyberliteracy is certainly a function of the ubiquity of the personal computer both at work and at home. According to the October 1993 Current Population Survey conducted by the U.S. Census Bureau, 23 percent of U.S. households had a home computer in 1993. Computer industry observers estimate that by the year 2000 approximately eighty percent of the households will own home computers, thus making the computer nearly as common in the home as a telephone. Use of high-powered modems and sophisticated communications software makes it possible for most home users to access general purpose databases for travel or home shopping information. Much of this information is provided in limited full text. In addition, school children have access to educational databanks that help them in their school work. However, current research is targeting comprehensive full text delivery including original graphics and images. Information professionals should be aware of the interesting, general purpose cybersources on electronic bulletin boards, local "freenets," and user interest groups on the Internet. Users will soon demand from libraries and information centers similar services (such as full text) presently supplied by popular sources in cyberspace. In order to connect to most cybersources a general user has to incur subscription and telecommunications charges. For people in dire need of information, but with limited means, libraries and information centers are likely to remain the institutions people fall back on for access to cybersources. They must be prepared to play that role.

4. The trend towards increasing privatization and ultimate commercialization of public information is also having a tremendous effect on the U.S. information arena. Some of the government information which used to be available in hard copy is being repackaged and sold online by information processing companies. Federal movements, such as re-inventing government, downsizing, and reduction of "big government," have accentuated this trend.

   The Census Bureau is the largest statistical agency in the U.S. federal government. It provides a very good example of federal initiatives to generate and distribute federal information in electronic formats. In 1984, CENDATA (an online service produced by the Census Bureau) was made available through Dialog Information Services and the Glimpse Corporation. Donnelly Demographics goes further by providing more detailed information based on data from the Census Bureau and can be accessed through the other services. While such electronic products provide highly accessible data, they are provided via commercial companies at relatively high costs. They are less accessible to individuals who have neither the money nor the technology to connect to the commercial organizations. Many of the 1990 census reports containing cross tabulations of data on geographic units can only be obtained on disc, magnetic tape or on disk. Prospective users who might want to access hard copies are thus eliminated by such a policy. Many other federal agencies are following similar policy with regard to their publications.

5. Current and dynamically developing PC technology is making it easy to download or upload large full text files. The use of sound cards; increasing memory and hard disk storage; increasing density of projection on display screens that results in better focus and higher resolution; increasing sizes of monitors and the accompanying video RAMs are laying the ground work for PCs to become true multimedia devices. The versatility of the resulting PC will enable the user to download and upload almost any cybersource with ease. The end user benefits from the increased functionality of the PC as most sources become accessible whether they are sound, image, text, or multimedia. Full text will indeed be full text as it is intended to be.

6. The portability of the current crop of PCs is redefining online access. Cellular telecommunications technology has introduced the concept of online services access without the line. With cellular technology, notebooks and personal digital assistants are now capable of accessing large remote cybersources independent of local telecommunications hard wiring. This mode of access to full text cybersources may have three implications for libraries and information centers. In the first place, when traveling out of town, libraries can be used as the home base access nodes to cybersources and assume the collective subscription costs for the client. Of course, libraries may completely be bypassed by affluent users who have direct access. Ideally, clients would be afforded a mix of options whereby the library or information center access is used for more specialized somewhat expensive cybersources, while direct access is reserved for free networks and electronic bulletin boards via the Internet. In the cyberspace lingo this "soft" connection, as contrasted to a modified form of "hard wiring," is used with complete transparency of both telecommunications and user gateway or bridge protocols and user interfaces. In other words, the user need not understand the underlying enabling technologies. In addition, the demand for "anonymous" user access will increase as traveling scholars and business executives graduate from simple information on demand to information on demand while in transit. De-

pending on the national information infrastructure policy that develops by the twenty first century, subsidies to local access centers should make this universal full text access possible. As potential access points, libraries and information centers must be designed with universal access in mind. This is particularly true of locally loaded cybersources, whether leased from external vendors or internally developed. The PC's remote access power is facilitated by the telecommunications connections discussed in the following section.

Fiber Optic Telecommunications Networks

There has been a quiet revolution in the telecommunications arena in the last two decades during which the conventional transmission, copper-wire medium has been replaced by fiber optic cabling. This revolution has laid the basic communications foundation that enables full text delivery to be implemented. Chapter two discusses telecommunications in the last quarter of the twentieth century in detail. This section summarizes the trend and relates it to full text information delivery systems. Full text transmission, which includes pictures and graphics, requires a very large volume of data signals to be accommodated by the telecommunications networks. In addition, very high speeds are needed so that images transmitted can be displayed at the receiving end in a timely manner. Inadequate transmission speeds slow the regeneration of images often resulting in unclear images. The characteristic that best measures the capacity of a medium for transmission of electronic signals is the size of the bandwidth. Transmission channels with broad bandwidth carry more signals than those with narrow bandwidth. If the material to transmit is full text, which contains images and sound as discussed in much of this text, broad bandwidth channels are ideal as images require more signals for transmission than plain text. Of the possible available transmission media currently marketed, fiber optic cables have the best combination of characteristics. Though small in size, they carry very large amounts of signals compared to twisted copper wires or coaxial cables. In the realm of full text transmission, optic cabling offers the most optimum transmission mode.

In the last 20 years or so, the major telecommunications carriers have invested billions of dollars to develop coast-to-coast fiber optic dominated communications networks supplemented by satellites and in some cases microwave radio links. AT&T, MCI,

GTE Sprint, and Western Union have been some of the major companies investing in this technology. Consequently, high speed fiber optic digital lines—T1 up to T4—are now crisscrossing the continental United States and are used daily to transmit full text information. In addition to the T series of digital data transmission lines, telecommunications companies have instituted digital PBXs and other signal switching devices which replaced electromechanical switching devices. Digital switching systems are now connected by digital transmission systems. This combination has made for effective transmission of data, including images.

At the international level, intercontinental underwater transocean fiber optic cables have been laid for the purpose of high-volume data communications. The use of the fiber optic cable is due to its attractive signal transmission characteristics. Since it has a low attenuation rate—the rate at which a signal loses strength at a given distance—repeaters are needed less frequently. Repeaters are used to regenerate signal strength at specified intervals. The fiber optic cable's large bandwidth is needed for graphics and images. Since it has virtually no electromagnetic interference, the optic cable has a low bit transmission error rate. This makes it ideal for alphanumeric text files typically used in full text cybersources. An example of underwater optic fiber cables used for digital signals is TAT-8 which connects the United States to Western Europe. It is owned by an international consortium led by AT&T, British Telecoms, and DGT (France). TAT-8 became operational in 1988 and is used by several major international companies, governments, and other organizations. Transocean fiber cables are supplemented by land-based fiber optic cables, satellites, and microwave transmissions to facilitate full text information delivery in cyberspace. The array of fiber dominated telecommunications networks currently being developed are ideal for digitally stored images, which require faster transmission speeds and larger bandwidth.

Electronic Digital Imaging

Digital imaging may be defined as the scanning and storage of documents in digital form as opposed to analog. Computer on microfilm (COM) library catalogs as well as other documents on microfilm are generated by a computer process using an analog mode. In the analog mode, the documents cannot be edited or otherwise manipulated. On the other hand, the digital format allows scanned documents to be edited if one has the appropriate software. The advantage of digital imaging is that scanned images can be stored on regular computers, searched by appropriate data base management systems (DBMSs) and transmitted to other remote computers using the conventional digital data communications channels. Research aimed at generating full text digitized cybersources, replete with graphics and images, is concentrated among large academic institutions, a few commercial companies, and Federal Agencies. The Library of Congress and the National Library of Medicine are among the federal agencies doing significant research in this field. UMI International and Eastman Kodak companies are among the leading commercial companies, while Cornell and Yale are among the academic institutions conducting high level image digitization research.

Emerging from these research efforts are fully fledged full text digital cybersources which can be searched remotely if one has the appropriate software. The sources can be browsed and downloaded if the user so wishes. For effective universal remote access to these databanks, five technical prerequisites must be fulfilled with regard to the scanned images of documents.

First, the national backbone (originally NSFNET, see Chapter Seven) that connects regional networks must have enough bandwidth to efficiently transmit the documents. Second, digital switching devices must be in place in the public telecommunications systems to speed the transmission of large amounts of data. Third, campus networks must be able to accommodate high-volume transmission. Fourth, there must be appropriate interface software to provide the end-user interactive access to the databanks. Finally, the access device (which in many cases is the personal computer) must have adequate capability to facilitate access, manipulation, and downloading of cybersource elements. Other than the national backbone, these other features have not been proven fully reliable. Therefore, the demand remains very high for practical systems that can deliver full text. It is conceivable that by the year 2000, viable systems will have evolved from the current prototypes. Documents on the World Wide Web (WWW) are similar to those envisaged. One should, however, be able to access and manipulate documents on or off the web. Image databases which are digitally encoded are just one of the variety of full text cybersources discussed in the next section.

From Online Databases to Cybersources!

Is the term database adequate and appropriate to describe the multitude and variety of information sources in cyberspace? How about the new sources found on the Internet's free nets and electronic bulletin boards concerning information? Furthermore, there are cybersources containing information on digitized illustrations and photographs. Should these be treated as databases comparable to the conventional bibliographic or full text databases? The database industry is changing quickly from its inception two decades ago. Back then it was easy to identify the major database vendors. The roles of the intermediary and the end users were distinct and fairly well defined. The categories of databases, as well as the communications modes used to access them, were also easy to isolate. The term cybersource appears to be a better candidate than database to describe the current sources within the changing online environment. The term embraces new sources not defined by the conventional term database.

Familiar database characteristics are gradually being metamorphosed. Many of the barriers, or lines of demarcation, are disappearing. Attempts are made every year by analysts in the industry journals and the *Gale Directory of Databases* to explain trends in content, vendors, and users. The most consistent feature noted by all analysts appears to be change! In lieu of more suitable categories, I will use traditional names to indicate the growth of databases in terms of volume. While I use the traditional categories as they appear in the published directories, I am cognizant of the multiplicity in subject content, source, format, and targeted end user. Organizations and individuals have created information sources on the Internet, such as the home pages or anonymous FTP (file transfer protocol) sites. Information in these sources may provide data as legitimate as standard databases such as ERIC or Medline. Since growth rates quoted in standard directories such the *Gale Directory of Databases* are based on traditional categorization, perhaps the volume and growth rates should be increased by five or ten percent to accommodate the nontraditional sources that provide database-like information in cyberspace.

According to the *Gale Directory of Databases*, there were over 5,300 databases by the end of 1993. The directory covers international online databases. As depicted in Figure 1.2, databases grew steadily over the last decade. One of the most significant trends is the growth of full text databases. Statistical sources indicate that

Online Databases 1984–1994

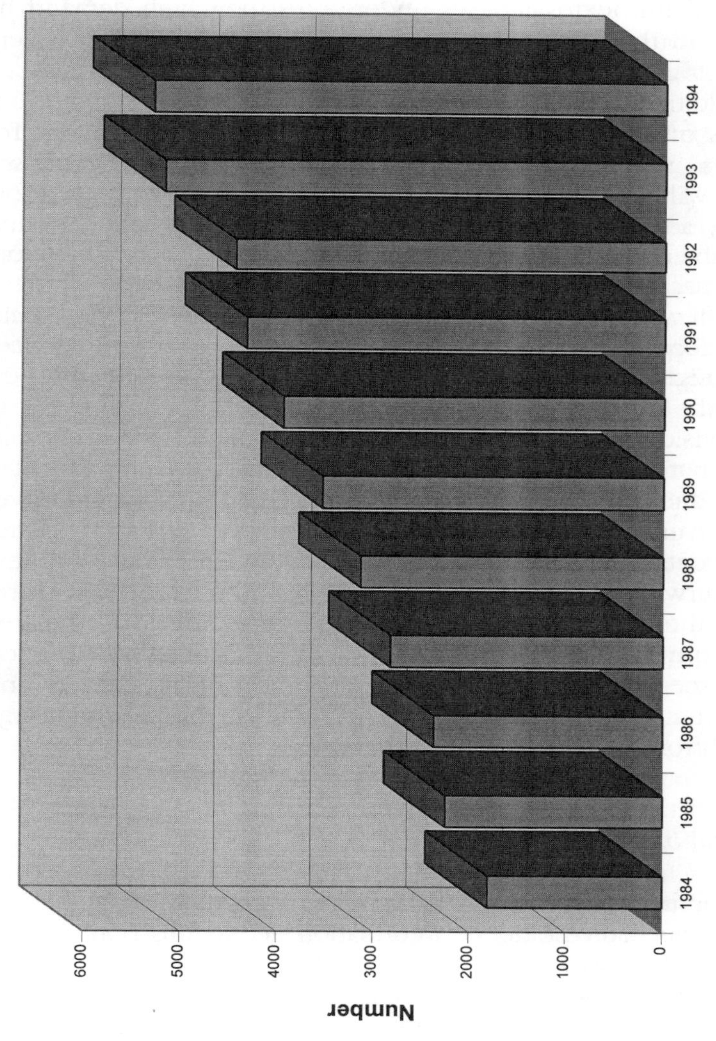

Figure 1.2

full text databases have grown 490 percent over the last ten years. This is not surprising since real life information clients would ordinarily prefer full text to bibliographic citations. The latter method adds the final step of locating the actual material as surrogated in the abbreviated bibliographic citation record. The tremendous growth of full text databases underscores their high demand by clients and the significance of developing user-friendly delivery systems.

In addition to full text databases, there are other significant changes on the database scene. In her "State of Databases Today:1994" report, published in the *Gale Directory of Databases*, Martha Williams reported that the number of database vendors grew by a factor of 15.5 (from 105 to 1,629) between 1975 and 1993. Revenue has increased from $40 million to over $1 billion. As for market share, three vendors have had the lions share in the last few years: Mead Data Central's LEXIS/NEXIS; West Publishing Corporation's WESTLAW; and Dialog Information Services (now Knight-Ridder Information, Inc.). The trio controls 85 percent of the use and 93 percent of the revenues.

In considering the totality of information and how it contributes to national development, we may consider using the concept of the Gross Information Product (GIP). This would cover all information services and products. GIP would include both traditional databases and what I have called cybersources—full text and otherwise. What needs to be factored into the GIP, with regard to the database industry, are the cybersources on the free nets, anonymous FTP sites, home pages, and bulletin boards on the Internet. Such sources, which are increasing in quantity and variety, tend to blur the analysis of connect hours and usage gleaned from the conventional databases.

Evolution of the End User

Most information users for whom libraries and information centers are the main points of information access are both literate and formally educated. While information professionals espouse the use of the Internet, they must recognize the fact that there will continue to be a sizable fraction of the public who will not use the information superhighway by the year 2000, but continue to use conventional sources in libraries and information centers. In other words, as we get carried away by the glamour of the

Internet, we must leave room in our information provision for users who may still want access to the traditional information sources. However, the most frequent library user is likely to have been exposed to the Internet. Currently, over 20 million people worldwide use the Internet. Industry observers estimate that by the turn of the century, that number will increase tenfold.

The needs of the database user, whether on the Internet or not, are becoming more varied. Three basic features will dictate user response to cybersource access. First, ease of use (especially for the nontechnical user) determines whether he or she will be a "repeat," and ultimately, frequent user. Second, timeliness of the information sought has a great bearing on repeat usage. Finally, accuracy of the information received will affect frequency of usage for information sources. Other user propensities will be centered around these three concepts. The client or information seeker in the foreseeable future will demand user-friendly interfaces to make seamless access to information throughout the world. Libraries and information centers must redefine their role in cyberspace as they provide the means and know-how of client access to virtual cybersources. As they retool, how can they contend with this variety of alternative sources? The global information village which will be both a blessing and a bane to the information professional, is discussed in the next section.

The Global Village in Cyberspace

Instantaneous virtual communities have generated a global cyberculture using the Internet as the link. Full text information delivery is promoting and enhancing communication in the virtual villages, as participants may have access to original material rather than indexes. User interest groups have been created that have virtual meetings, including most of the nuances of community life such as intimacy, agreements, conflicts, conflict resolution, disagreements, and ultimate splits. In addition, the virtual villages have their share of deviants, who would rather not obey the rules, downplaying what others regard as significant. More destructive elements are not missing either: Hackers plant viruses on the network for the fun of it. Six major implications are apparent for information specialists.

First, ease of access to alternate sources relieves the library or information center of the burden of being the sole or even the major source of information in a given community. With so much material available worldwide, both the client and the information specialist are at an advantage. Client frustration is reduced as alternative sources provide material which the host institution might not have, let alone afford to purchase. The Internet facilitates information sharing across national, ethnic, religious, and many other barriers. Linguistic barriers remain significant. While it might be fascinating for a user in New York City to connect to an online library catalog in Venezuela, the excitement stops when instructions to use it are in Spanish and the New Yorker cannot translate!

Second, the information specialist has access to a broader sweep of information sources, in addition to the traditional ones. This is a bane in that he or she has to learn new tools. It is, however, an advantage to access sources that would not otherwise be available for use because the parent or host institution cannot afford them. A combination of effective interlibrary loan identification services, electronic document delivery, and electronic mail (e-mail) have improved the services that libraries and information centers provide.

Third, the library or information center is progressively facing tough competition over information provision. Given powerful telecommunications systems, powerful multimedia computers for accessing remote cybersources, and a wealth of external sources, the client will most likely prefer to access external material unless material provided by the host is free or superior in quality. The "bypass syndrome" will affect many libraries and information centers unless they have unique services to offer the client. A major obstacle is the cost of information access. Several sites on the Internet provide samples that lure the user to visit before they realize that the substantive information can only be obtained through subscription.

Fourth, in terms of budgetary allocations and resource acquisitions, the amount, level, and intensity of decisions also requires different approaches. For instance, if a U.S. user can get access to information about the Australian economy directly from The National Library of Australia, the parent institution will most likely minimize collection of material on this subject.

Finally, skills upgrade for information specialists becomes mandatory for survival. While traditional skills can be updated with little or no computer skills, proficiency at navigating the Internet does require a considerable amount of computer literacy. In other words, full exploitation of the global village by information professionals for client services needs more computer skills than the current crop of information professionals possess.

THE AGE OF INFORMATION ON DEMAND

The concept of information on demand has two connotations. First, from the information user's point of view, information is provided to support decisions that are of current and immediate concern to them. This implies timeliness, relevance, and accuracy. Second, from the information provider's point of view, information is identified and acquired at the time of request. This implies that the library or information center need not have the information sources on its premises as long as it can have access to them when demand is triggered by a user's request. Both connotations call for providing usable information rather than providing directions to sources. Although indexes are needed to facilitate selection of relevant sources, availability of full text is imperative for a user that needs a deeper coverage of the subject.

While there are several functions that libraries and information centers provide to their clientele—such as reading rooms, meeting or conference space, information provision is paramount among them. Four main factors interplay to highlight the significance of information on demand. First and foremost is fiscal crisis. Libraries and information centers are going through lean times in terms of budgets. Many of them are cutting down on services and materials to purchase. At the same time, costs especially for journal subscriptions, are spiraling upwards. Most libraries have had no alternative but to depart from expanding holdings, thus providing access to information sources and acquiring information on demand.

Second, in economic terms, real demand results in services. Technology has consequently responded to the high demand for information analysis, storage, and dissemination. Many of the cyber trends that involve technology have been bolstered by information on demand. The PC had to grow up in order to handle the sophisticated full text files. Image databases and telecommunications are part of the infrastructure needed to equip cyberspace for full text delivery.

Third, business and industry comprise a sizable portion of the activities in the information arena. Business competition and strategy are based on getting the right information at the right time. Currently, leading companies are developing executive information systems (EIS) that supply executives with the most up-to-date information at their desks. Financial information, which may include currency values, stock prices, and futures, is a good example of a category that calls for up-to-the minute currency. PCs and telecommunications are the enabling technologies that have made the office computer a true workstation with remote access to databanks. The workstation designation of a PC implies a multifunction personal computer, which may be used to: draft reports; do statistical analysis; send and receive electronic mail; and send and receive faxes. In addition, it can download full text information, massage it in spreadsheets and include it in corporate reports.

Finally, international competition for business has made information on demand the sine qua non of information services. Both small and large U.S. companies are doing business overseas. The timeliness and accuracy of information may make or break an international contract. Information systems that can provide full text rather than citation are thus prime sources in the international business arena.

THE USE OF CABLE MODEMS FOR MULTIMEDIA COMMUNICATIONS

Four factors have prompted CATV companies into providing diversified services that have extended into data and multimedia. First, experiments in interactive television have given the companies confidence that two way communications modes are feasible on their networks. Second, the 1996 Telecommunications Act removed restrictions on CATV providers' participation in data communications and information services. Third, the search for services to provide in the information market beyond cable TV serves as a catalyst to develop consumer oriented products. Finally, the customer need for bandwidth for Internet access has made CATV networks possible alternatives.

Cable modems are somewhat a misnomer as they do not do the same functions as the conventional telephone line based modems. They are devices that modify the regular CATV cable circuits to transmit signals in a bi-directional mode. The main advantage of cable modem use is the bandwidth which may be 3–10 Mb/s. Technology improvements are being made and this figure is likely to treble in the next few years. The prime target markets would include the home general user for Internet access; the telecommuter who must access, download, and upload a variety of information products from the corporate headquarters computer system; and company to company transactions. Libraries and information centers should investigate possible use of cable modems for resource sharing, especially with regard to interlibrary loans. With digital image collections currently being developed, this medium would be a potential choice. Cable models place small companies on par with large organizations that have T1–T3 leased lines.

References

Albiges, Luce-Marie. "Remote Public Access to Picture Databanks." *Ifla Journal* 18 (1992): 37–42.

Arrigona, Daniel. "Federal Business Information in Electronic Formats." *RQ* 27 (1988): 322–333.

Brockman, John R. "Information Management and Corporate Total Quality." *Journal of Information Science* 19 (1992): 259–266.

"Computer Use is Growing." *Census User News—Section 05.* June 21, 1995.

Gale Directory of Databases, Volume 1: Online Databases July 1994, Edited by Kathleen Young Marcaccio. Washington, D.C.: Gale Research Inc., 1994.

Jackson, A.E. "A Retail Opportunity in the Contemporary Information Market." *Aslib Proceedings* 44, no. 10 (October 1992): 341–349.

Keiser, Bernard E. and Eugene Strange. *Digital Telephony and Network Integration.* New York: Van Nostrand, 1984.

Keyes, Jessica. *Infotrends: The Competitive Use of Information.* New York: McGraw-Hill, 1993.

Kibirige, Harry M. "Information Communications Highways in the 1990s: An Analysis of Their Potential Impact on Library Automation." *Information Technology and Libraries* 10, no.3 (September 1991): 172–184.

Kinnucan, Mark T. "Demand for Document Delivery and Interlibrary Loan in Academic Settings." *Library and Information Science Research* 15 (1993): 355–374.

Langenberg, Donald N. "Information Technology and the University: Integrated Strategies for the Twenty First Century." *Journal of the American Society for Information Science.* 45, no. 6 (1994): 323–325.

Lievrouw, Leah A. "Information Resources and Democracy: Understanding the Paradox." *Journal of the American Society for Information Science* 45, no. 6 (1994): 350–356.

Marchand, Donald A. "Infotrends: A 1990s Outlook on Strategic Information Management." *Information Management Review* 5, no. 4 (1991): 23–32.

Morris, Ruth C.T. "Toward a User-Centered Information Service." *Journal of the American Society for Information Science* 45, no. 1 (1994): 20–30.

Offer, Colin. "Charging for Information: A Changing Climate and an Emerging Market Place." *Informediary* 5, no. 1 (1991): 49–62.

Pugesley, William and Rickyo Lennon. "A Report on the Status of Fee-Based Information Brokering in the United States." *Informediary* 4, no. 1 (1990): 13–32.

2. Cyberspace's Foundation: Telecommunications

A major challenge facing information professionals and their clients is the unlimited access to information sources in cyberspace at the national and global levels. The phrases "information explosion," "information overload," or "information glut" are hardly sufficient to describe the phenomenon. Perhaps information superflow may be substituted. With one keystroke, a tremendous amount of information is literally thrust at your face on the computer screen in an instant. If one has the appropriate computer system and software, technology is available for accessing, receiving, manipulating, and downloading full text information at ease. Such information is replete with graphics, images, text, sound, and animation. It is almost magical, stupendous, and unimaginable. While clients may not care how it got there, a sizeable number of professionals and their clients are beginning to demand user friendly explanations of how one gets a digitized picture, a large piece of text, or sound on a computer system from coast to coast within the United States. Furthermore, one would wonder and wants to know how such transmission is possible on an intercontinental basis. This chapter provides as nontechnical an explanation as possible of how telecommunications technology facilitates full text information delivery systems. Refer to references at the end of this chapter or the selected bibliography at the end of the book for further information.

TELECOMMUNICATIONS 101 SYNPOSIS

Telecommunications is the technology that facilitates transmission of information from one point to another by electronic or electromagnetic means. The Greek prefix "tele" stands for distant and the suffix "communications" means giving or exchanging information, signals, or messages. There must be a source of information (or the sender), a transport mechanism, and a target or receiver. In telecommunications terms we have the transmitter, transmission line, and receiver that compose elements of a generic basic telecommunications system (see figure 2.1). As for equipment, three sets are needed: a transmitting device(s), transport devices, and a receiving device(s). To move from one place to another, information goes thorough six possible stages: input, display, encoding, transmission, decoding, and display (see figure 2.2). The first three stages are performed by the transmitting devices, the fourth by the transport devices, and the last two by the receiving devices. In telecommunications, a basic problem arose early in the history of the technology: spoken language as a medium of communication. Language is too ambiguous for machine operation. In order to speed up transmission of information via electronic devices, it had to be coded into electronic signals that machines would understand and thus effectively execute a communication session. Coding introduced precision and efficiency in transmitting human messages. The process is reversed at the receiving end from machine language to human natural language in order to make the message intelligible to the human operator or communicator. Over the years, a number of methods have been invented to perfect and streamline signal transmission.

Analog and Digital Signals

Telecommunications technology uses two main types of code: analog and digital. In the analog method, the signals are in a continuous wave form varying in frequency and amplitude, whereas in a digital method, they are in discrete minuscule pulses with an on or off format as shown in figure 2.3. Historically, voice generated information has usually been characterized as analog and computer generated information characterized as digital. However, voice and data may be transmitted in either analog or digital forms.

Figure 2.1 Basic Telecommunications System

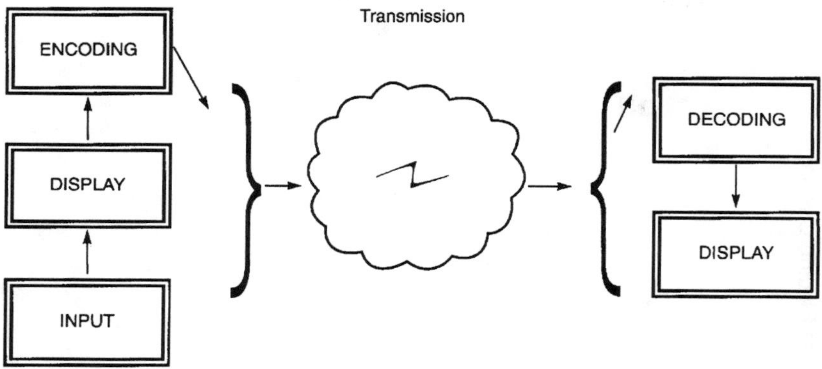

Figure 2.2 Information Flow in a Telecommunications Channel

For generations, most telephone switching systems were electromechanical, connecting analog transmission channels. As mentioned in Chapter One, several U.S. national telephone carriers have switched to digital switches and transmission lines. When transmitted over long distances, both analog and digital signals lose strength, a characteristic known as the attenuation rate, and have to be boosted. In analog systems boosting is done by amplifiers to sustain signal strength, while digital systems use repeaters. The main disadvantage of analog systems is that amplification includes amplifying extraneous noise and the final data may be corrupted. On the other hand, repeaters do not pick up noise as they regenerate the signals and thus result in more precise and accurate transmission.

Bandwidth

Transmission channels (or transmission lines) and media have different capacities for communicating information or signals. Depending on the user requirements and characteristics of the media used, various capacities may be designed for a telecommunications system. As transmission media, copper wire, coaxial cable, and optic fiber have varying capacities for carrying signals. Copper wire has the lowest capacity and optic fiber the highest, respectively. The term bandwidth is used to describe the trans-

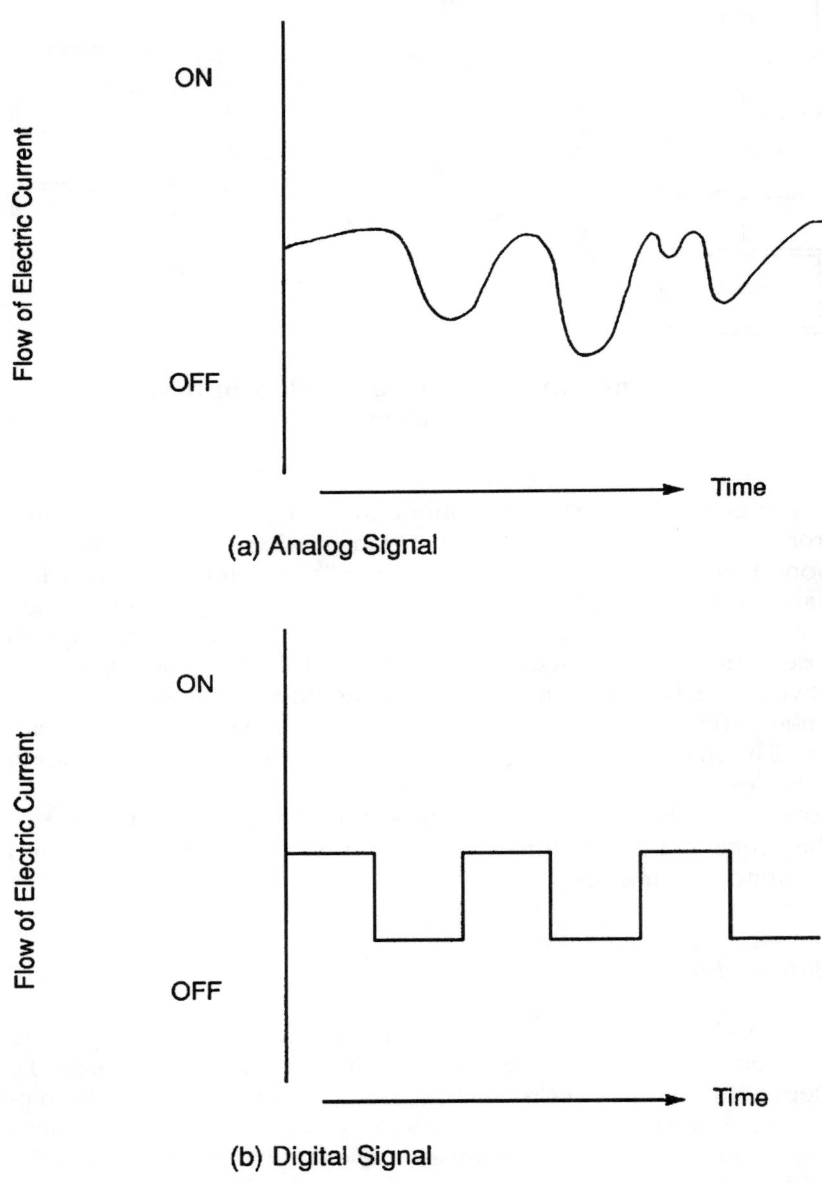

Figure 2.3 Analog and Digital Signals

mission capacity of a channel or medium. It refers to the range of frequencies a channel can transmit. In mathematical terms it is the difference between the highest and the lowest frequency. It is ordinarily quoted in hertz which is the equivalent of cycles per second. It may be expressed in kilohertz (kHz), thousands of hertz per second or megahertz (mHz), millions of hertz per second. A regular analog telephone line transmits at approximately 3 kHz which is between 300 and 3,400 kHz. This is the standard range at which much of human voice communication is carried over the common analog telephone.

In practical terms, the terms high or low are used with regard to bandwidth to describe the ability of a transmission channel or line to carry information or signals. Depending on the way it is designed or the basic characteristic of the media it uses, a given channel may have either high or low bandwidth. The higher the bandwidth, the higher the rate at which a channel can carry information. Certain applications require more signals at a faster speed to operate effectively. Full text information delivery containing images and graphics requires more bandwidth than plain text. Images and graphics start with a large amount of signals (bits) to represent them in the machine-encoded language. In order to transmit them fast enough to be replicated properly at the receiving end, the transport channel must have the bandwidth to accommodate the "burst" of signals.

Serial and Parallel Transmission

The computers used when preparing information for communication store, or organize, the data internally, character by character, in groups of bits. Ordinarily, the information is stored in groups of seven or eight bits (7 or 8 bits making a character—byte). The eight bits that make a character (byte) may be designed to travel in parallel circuits along the channel simultaneously to other computers or to high-speed printers. This mode of transmission is called parallel transmission. It is usually faster and accommodates high-speed devices. The devices that make the connection possible are called parallel interfaces.

In some cases external devices, especially modems, may only receive signals one bit at a time. In this case the seven or eight bits that compose a byte must be transmitted one at a time over the channel. When such a mode of transmission is used it is said to be serial transmission. A serial interface is used to connect the computer to the modem. In the personal computer market, communication boards or COM ports, are used to convert from parallel to serial. Depending on the make of the PC, two or more COM ports may be available to connect to external devices.

Synchronous and Asynchronous Transmission

In a telecommunications session, communicating devices must match their modes of signal transmission—a process called synchronization. This is done to monitor messages passed between the sender and the receiver to make sure that the message sent was received intact, as it was intended. Two methods are used to achieve synchronization, namely asynchronous and synchronous. They were devised to accommodate interactive, real-time, human-oriented access (asynchronous) and machine-to-machine access (synchronous).

Asynchronous transmission is character based in that the transmitter sends one character to the receiver at a time over the telecommunications channel. Each character is preceded by, and ends with, start and stop signals (bits). Asynchronous transmission generally originates from slow-speed terminals. An operator keys in characters and the system sends them over to the receiver one at a time to compose the message. The method is slow and wasteful of computer and communication time. It is, however, useful in interactive, real-time applications where dumb terminals are used or an intelligent terminal's storage capability is not utilized.

Synchronous transmission, on the other hand, is based on blocks of characters. Transmission control signals called "synch" signals are used to indicate the beginning or end of a block. Such signals are fewer in this mode of transmission than in the asynchronous mode and more messages can be sent at a given time, depending on the channel speed and device characteristics. Synchronous transmission is usually used for computer-to-computer applications or with terminals or processors that have cassettes or storage buffers. Blocks of characters are transmitted at high speeds resulting in efficient communication sessions. This mode of transmission permits transmission of large volumes of signals typical in full text information delivery systems incorporating images.

Simplex, Half Duplex, and Full Duplex Channels

Telecommunications channels (lines) may transmit in either one direction or two directions depending on their design (see figure 2.4). Two kinds of channels are used to transmit in both directions and one kind is limited to transmission in one direction. Within the North American telecommunications context, simplex channels transmit in one direction only; half duplex channels transmit in both directions, using one direction at a time; and full duplex channels transmit in both directions simultaneously.

A full duplex channel is equivalent to two simplex or half duplex channels and is usually referred to simply as duplex. The half duplex mode requires a pause at the end of each transmission for the receiver to respond, normally referred to as the turnaround time. The full duplex mode needs two channels simultaneously transmitting in different directions.

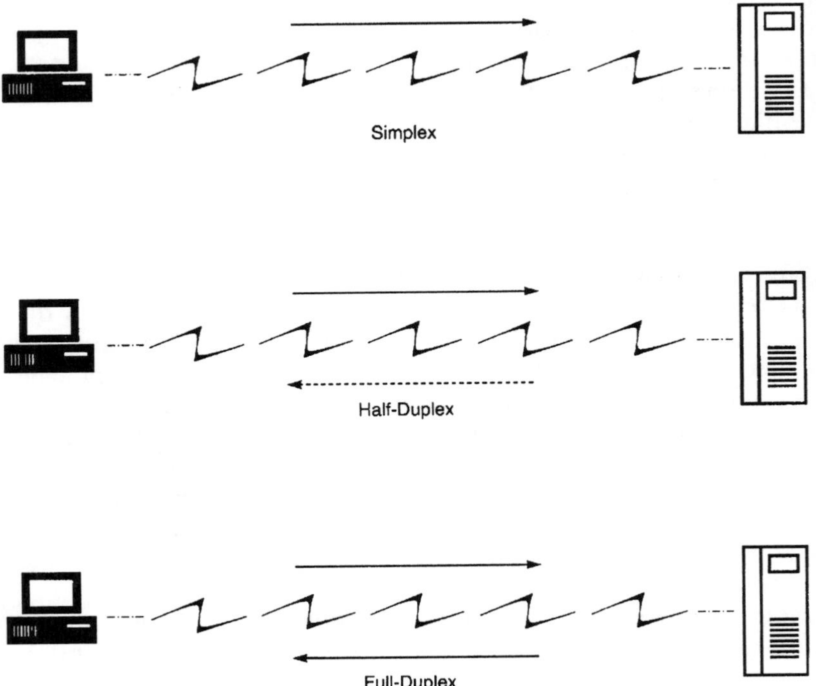

Figure 2.4 Modes of Transmission

In terms of wire requirements, simplex and half duplex require two wires to complete the necessary electronic circuit. Full duplex, on the other hand, ordinarily needs four wires to complete the requisite circuit. Transmission devices usually have precise specifications indicating wiring requirements—either two or four wires as the case may be.

Modulation, Demodulation, and Modem devices

Signal or information transmission over a telecommunications channel may be either analog or digital. There are methods and devices that convert one mode to the other. If a digital transport mechanism is used in the session, such as the AT&T Digital Data Service (DDS), digital-based devices that have to connect to it need a data service unit/channel service unit combination (DSU/CSU). The DSU is a synchronous loop discussed under networks dataline driver and the CSU deals with the local on pages 7–8. Unlike the analog to digital conversion, the connection devices do not alter the mode of transmission.

On the other hand, if the transport mechanism (channel) is analog, like many dial-up connections, a device is needed to translate the digital data generated by the computer. By doing so, the data stream generated by a computer terminal can be transmitted over the telephone line. Once the bandwidth of the carrier channel is identified, the conversion device generates one or more signal frequencies within the range of the channel bandwidth referred to as "carrier frequencies." So equipped, the device encodes the digital data and imposes it on the carrier (channel). This process is called modulation of the carrier signal. At the receiver end, the process is reversed by another conversion device and analog signals are demodulated into digital signals for the target computer device which the user can manipulate as needed. The term "modem" was derived from the process of modulation and demodulation. Naturally, in any computer dialog using analog channels for communication, there must be modems at both ends of the channel.

Modems have rapidly grown more sophisticated during the past decade (see figure 2.5). They can be inserted into the computer as extension boards within the slots provided by the computer manufacturer or connected externally via a serial port. Their speeds have increased immensely from 300 bits per second (bps) to over

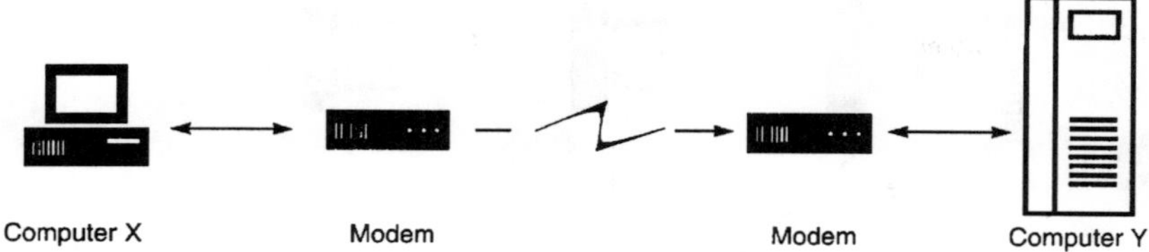

Figure 2.5 Modem Linked Transmission

28,880 bps. Coupled with communications software and powerful personal computers, full text documents with graphics, images, and sound can now be downloaded to office or home desktop computers much faster than ever before.

Modem speeds are measured in bits per second (bps). Since seven or eight bits make one byte (character), it is possible to estimate how many pages of text per hour a given modem may deliver. However, other considerations may complicate the estimate, including the extraneous bits attached to messages for error checking, the rate of the transmission channel and the intelligence of the receiving computer device, and whether the receiving computer has temporary storage buffers to capture incoming information signals. Baud rates measure the speed at which modems send or receive signals from remote computer devices. A signal may have more than one bit, thus one baud is not always equivalent to one bit. In other words, 2400 baud is not necessarily the same as 2400 bps—unless the transmission channel operates at a rate of one bit per signal. For modems in general, higher bps speeds are better suited for full text documents that include images and graphics.

Multiplexors

Multiplexors, sometimes abbreviated simply as mux, are devices used to combine signal input from a number of low-speed channels to be transmitted over one high-speed transmission line. They are often used when clusters of terminals send and receive information from a remote computer. The signals are captured from the individual terminals, stored in the multiplexor's temporary storage buffers, and sent as blocks of messages to the host computer. The process of multiplexing streamlines and optimizes signal flow on the line (see figure 2.6). Several methods for multiplexing are

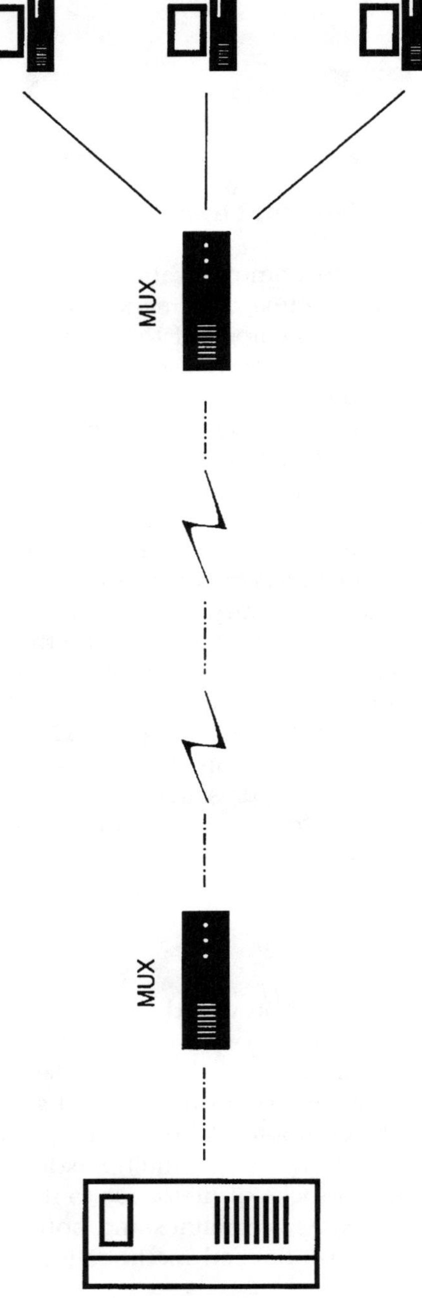

Figure 2.6 Multiplexor

used in the telecommunications industry. In the realm of full text information delivery, multiplexors allow several people to share a "host" computer that may contain common information. The host may house commonly used software or shareable databases. Multiplexors are vital to full text delivery systems, which inherently involve large volumes of signals, because they regulate signal flow from the host to terminals.

Transmission Media

The transmission medium is the physical link between two or more communication devices. Much of modern telecommunications is conducted on networks, but linked devices need not be part of a networking environment. The most common media include twisted pair copper wire, coaxial cable, optic fiber, microwave, and satellite.

Until relatively recently, twisted copper wire dominated the telecommunications industry as the medium of choice. This was partly because of its extensive use in transmitting voice communications over telephone lines. It still forms much of the backbone of the local telephone network. Two insulated wires are wound around each other to minimize interference. Early computer data communication used it since it was available as part of the telephone-installed base. Under normal circumstances, it can support signal communication up to 10 Mbps (million bits per second) over hundreds of meters. Its advantage is that it is easy to install, and for many organizations, it is already in place. Its main disadvantage is that it is susceptible to electromagnetic interference that results in distorted signals at the receiving end. This problem can be minimized by shielding the pair with a metallic alloy. Twisted pair copper wire is gradually being replaced by fiber optics.

Coaxial cable (coax) (see figure 2.7), is another telecommunications medium. It is composed of a central copper insulated wire, which carries the signal, surrounded by an insulating material often referred to as a "dielectric." A second conductor forms a sheath around the dielectric and is also shielded by an insulating material, usually plastic. Coaxial cable has a higher signal carrying capacity than twisted pair wire. It can carry signals ranging in frequency from 5 or 10 MHz to 400 MHz. Because of the way it is designed, coax requires fewer amplifiers because its rate of attenuation is lower than twisted pair wire. Its typical usage has

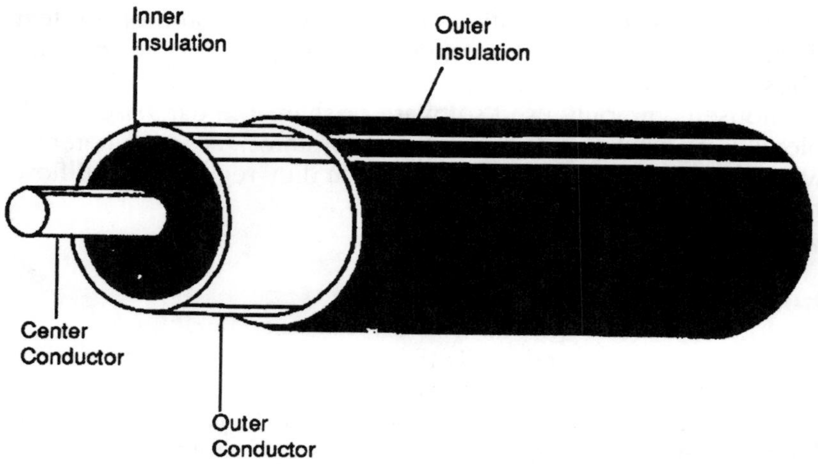

Figure 2.7 Coaxial Cable

been in connecting telephone company central offices (COs) and cable television networks. It is gradually being replaced by optic fiber as the preferred CO connecting medium.

Optic fiber cables are made from very thin glass or plastic filaments surrounded by a plastic jacket, or sheathing. There may be more than one filament in a given cable. Signals are transmitted by placing a controllable light source at the sending end and a light sensitive receptacle at the receiving end. Light sources used are usually lasers or light emitting diodes (LEDs) and the receptacle a "photodiode." Similar to regular modems which modulate carrier signals, an optic fiber modem controls the light source to facilitate signal transmission over the optic fiber.

Optic fiber has an immense bandwidth compared to other media. It may be in hundreds of millions or even billions of hertz (GHz) per second and can carry signals at billions of bits per second (Giga bits per second or Gbps). It is light in weight and smaller in size compared to copper cables. These characteristics combined with bandwidth and speed make it attain extremely high capacity per cross section or diameter of cable. Many conduits designed for electronic wiring in buildings, under city streets, or on utility poles are overcrowded. Optic fiber makes an exceptionally efficient alternative.

Optic fiber is virtually immune to electromagnetic interference in contrast to copper-based cables which experience noise-related signal distortion. In addition, it is very secure. Since it does not radiate electrical energy, it is difficult to tap by signal pirates. In the case of copper cables, tapping requires relatively simple contact with the metallic conducting material. Optic fiber tapping requires breaking and reconnecting it. Thus any tapping or attempts to divert signals will be noticed by the line administrator.

Three main disadvantages have restricted the universal deployment of optic fiber for communications. First, it is still more expensive than metallic cables. Second, splicing it in case of breakage requires very high precision, unlike other cables. Third, it is incompatible with the existing cabling. With the increased market, these disadvantages are gradually diminishing. Long distance lines constructed by U.S. national telephone carriers in the 1980s incorporate them. Use of optic fiber has so far been one of the most significant features of full text transmission within the last decade. The National Research Foundation's research network telecommunications backbone, the initial foundation of the Internet, is predominantly optic fiber.

The T series (i.e., T-1, T-1 c, T-2, T-3, T-4) are telecommunications standards that define long distance digital lines. AT&T and other North American telecommunications carriers own T series lines for leasing, many of which are fiber and digital. Their transmission speeds range from 1.544 to 274.176 megabits per second (Mps). Many institutional networks lease dedicated or shared-fiber, optic-based, digital T-series trunk lines from national carriers to form their proprietary information networks such as MANs and WANs. In addition, a number of universities have installed backbone fiber optic based networks for on campus, high-speed transmission. These trends augur well for full text transmission which calls for speed and high bandwidth.

Microwave is an alternative transmission medium based on high frequency radio waves. Signals are transmitted through the air between pairs of microwave antennas mounted on very high towers (see figure 2.8). There is a line-of-sight requirement: for the relay antennas to communicate, they must be in sight of each other (radio waves follow the earth's curvature, however, radio line-of-sight is longer than human visual line-of-sight). Relay antennas are placed at intervals, between thirty to fifty miles apart. The microwave systems may have a capacity of between three to twenty billion bits per second (Gbps). They are used for high-volume, high-speed, and relatively high-quality computer data communi-

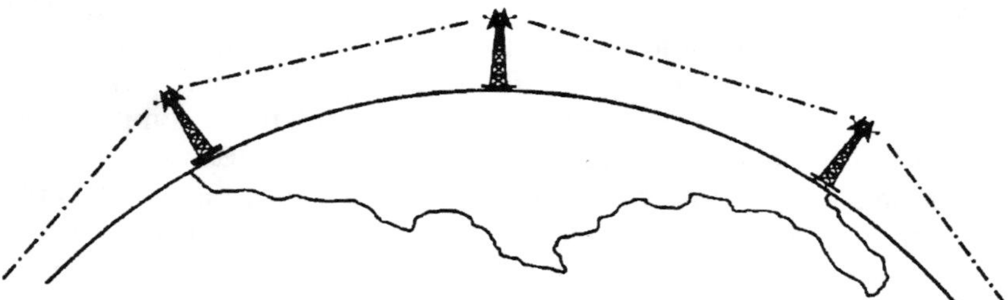

Figure 2.8 Microwave Relay Towers

cations. Due to their transmission capacity, microwave systems are suitable for full text applications that contain large amounts of text, images, and sound. There are pros and cons to microwave signal communication. The main disadvantage is the attenuation or distortion caused by atmospheric conditions. Very heavy rain, fog, or thunder, may affect microwave transmission. One of the greatest attractions of this medium is that microwave does not require physical right of way. In addition, installation does not involve digging trenches or pulling cables over utility poles. It is thus ideal for sparsely populated areas, areas with tough terrains, or areas where legal or proprietary issues make it difficult to lay regular cables.

Satellites are yet another alternative medium for telecommunications. Their mode of communication is similar to microwave in that they are based on high frequency radio waves. The transmission mechanism includes a communication satellite and two or more ground stations. Like microwave, they have very high bandwidth and do not need right of way. Satellite earth stations that are used for data communications are similar to dish antennas which are commonly used by individuals and organizations to receive television signals.

Most communication satellites are placed in what is called a geosynchronous or geostationary orbit—an orbit timed to the earth's rotation. With such an orbit, the satellite remains in a fixed position in relation to the earth's antennas. This obviates the need for constant reorientation of the earth stations in order to remain in touch with the communications satellite. A common distance for geostationary communications satellites is approximately 23,000 miles above the earth's surface (see figure 2.9). Satellites have several components that transmit and receive signals called "transpon-

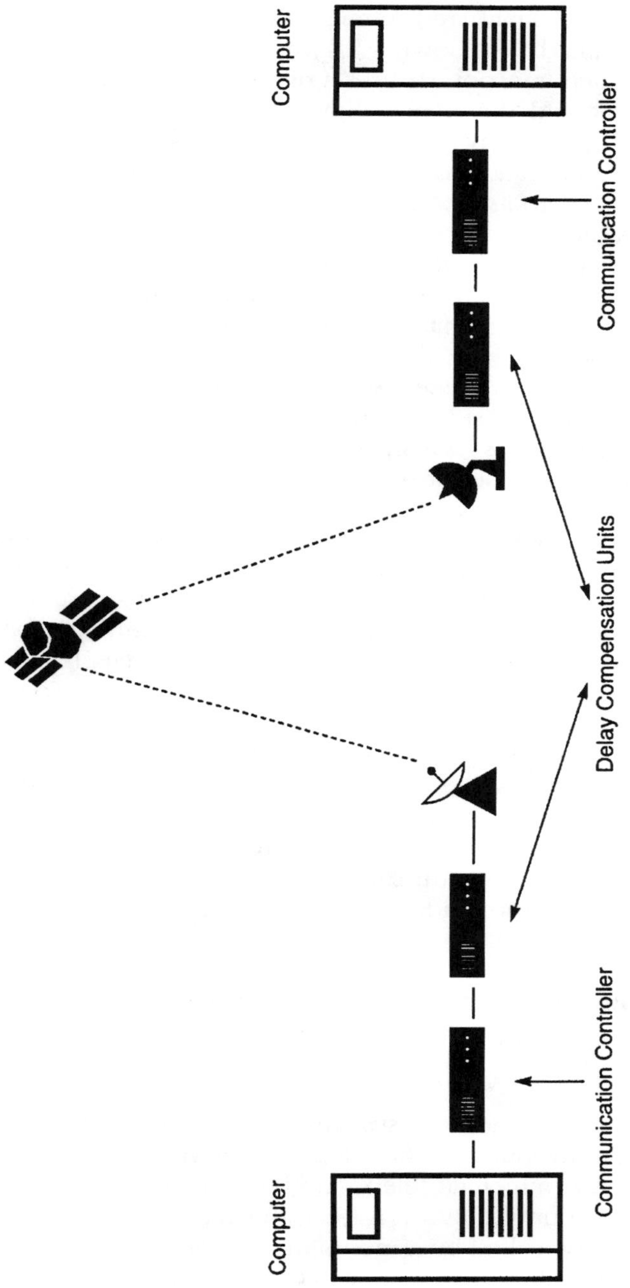

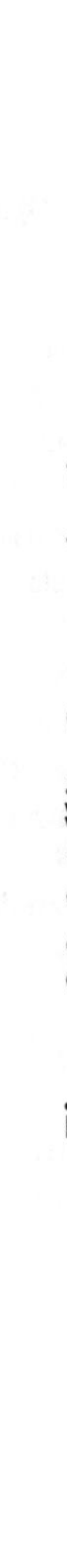

Figure 2.9 Satellite Communications

ders." A transponder receives signals on a channel called an "uplink," regenerates the signal, and retransmits to earth stations via another channel called the "downlink." To increase the number of transmissions, each transponder uses a different pair of frequencies for the uplink and downlink, allowing for several simultaneous transmissions at any given time. In terms of distance covered, each communication satellite has a geographic area which can receive its signals, to its "footprint." The term footprint is used to denote the maximum area on the earth's surface within which signals from a given space satellite may be received. Points beyond the footprint will not receive such signals. Earth stations within the footprint can transmit or receive signals to one another if appropriately tuned.

Among the advantages of satellite data communications is the broad bandwidth which is suitable for full text delivery systems. Their transmission mode is essentially broadcast in nature. This means that messages beamed to earth can be picked up by any station that is tuned to the transmitting downlink channel and pointed towards the communicating space satellite. While this allows the satellite to send signals to numerous earth stations simultaneously, within its footprint, privacy cannot be easily maintained. For corporate data or otherwise confidential data, scrambling or encryption is normally used. At the time of usage, such data has to be deciphered using special conversion devices. Another serious problem is the propagation delay caused by transmitting signals through space. All satellite signals using a relay station in ordinary geosynchronous orbit are subject to a quarter second delay in both directions. While the delay may be vital for real time interactive applications, file transfers can be done with relative convenience. Due to its very high bandwidth, satellite communication is very conducive to full text information delivery systems.

TRENDS IN TELECOMMUNICATIONS

Historical Perspective

Present day telecommunications systems and networks owe their origins to the 1876 invention of the telephone by Alexander Graham Bell. The first commercial telephone became operational in 1877. From that date until January 1 1984, the U.S. telecommunications scene was controlled by a monolithic telephone empire—

Bell Telephone—managing both local and long distance phone services. The system was composed of the parent company, American Telephone and Telegraph (AT&T), Western Electric, Bell Labs, and local Bell telephone companies. It took a century of operation, antitrust suits, and Judge Harold Green's decisive ruling to bring AT&T's sole monopoly on the telecommunications business in the U.S. to an end on January 1, 1984.

The telecommunications networks as we know them today have sprung from relatively recent regulatory measures. The AT&T divestiture on January 1, 1984 created seven Bell operating companies (dubbed the "Baby Bells") as follows:

- Ameritech
- Bell Atlantic
- Bell South
- Nynex
- Pacific Bell
- South Western Bell
- U.S. West

A common research organization was maintained in the name of Bellcore which emanated from the old AT&T Bell Labs. The parent company, which won the nickname Ma Bell, retained the long distance services, while the Baby Bells concentrated on the local markets. The ruling successfully opened up the telecommunications industry to other companies in the past decade. Long distance companies like MCI and Sprint have successfully competed with AT&T in delivering long distance telecommunications services.

Divestiture of AT&T brought more than new administrative organizations to the telecommunications scene. New products and services were introduced, such as intelligent handsets and desk equipment. Among innovative services, MCI was one of the pioneers in introducing electronic mail (e-Mail) for general public use among its offers of services. E-mail has become one of the main methods of transmitting full text documents on the Internet and other networks.

Procedural Changes in Information Processing

Organizational changes that stripped AT&T of its unchallenged supremacy occurred concurrently with procedural changes. While both changes were technology driven, procedural changes were more affected by the emerging software, computer hardware, and telecommunications developments. Three processing phases can be isolated that required new telecommunications systems.

First, the 1960s marked the era of time sharing, when large computers dominated the scene. Most users had hard-wired connections to mainframes that they used via computer labs. Dial-up services were virtually unknown, and information exchange among computers was done by proprietary protocols, such as IBM's System Network Architecture (SNA).

Second, the 1970s and early 1980s ushered in the era of distributed processing, where more than one site was used to process data. This lessened the burden on the mainframes, eased the congestion at the main computer centers, and gave remote stations options in the final processing and displaying of the data. Telecommunications companies scrambled to provide the links—especially in the late 1980s.

Third, the mid to late 1980s witnessed an unprecedented growth in networks. The U.S. data processing environment was strewn with local area networks (LANs), metropolitan area networks (MANs) and wide area networks (WANs) (see Chapter Four). The Transmission Control Protocol/Internet Protocol (TCP/IP) networking standard developed by the U.S. Department of Defense became the de facto networking standard for connecting disparate machines for the purposes of exchanging information. Phase three of the procedural developments, has thus made it possible, easier, and effective to send full text from computer to computer.

Individuals and organizations currently exchange full text information via networks, such as LANs, MANs, and WANs. Using the basic telecommunications elements discussed in this chapter, coast-to-coast transmission may be implemented by a combination of fiber optics, microwave, satellite, and twisted pair copper wire as depicted in figure 2.10. The Integrated Services Digital Network (ISDN) promises to make the networking effort more effective in full text information delivery.

The ISDN Factor

ISDN was discussed in some of my earlier works as a promising technology that will soon enhance information delivery and exchange (see figure 2.11). I was among many industry observers that made that prediction, judging by the number of industrial enterprises with stakes in it. In the last decade, ISDN has had

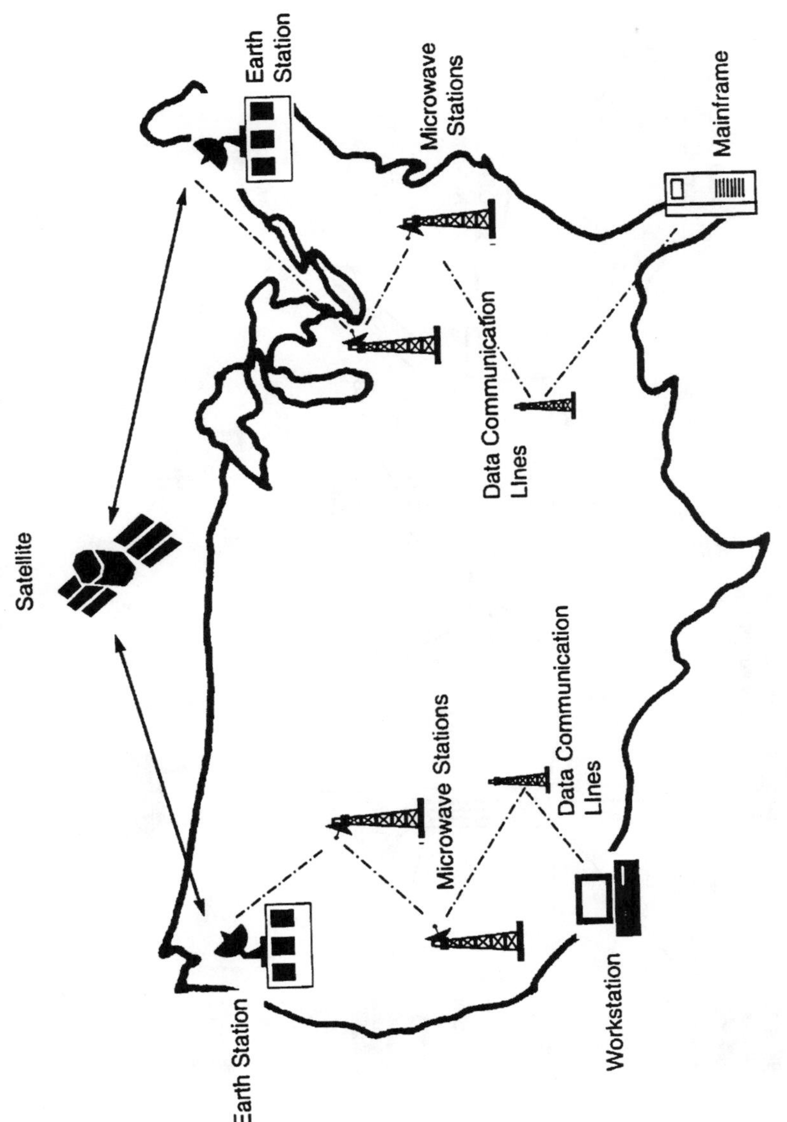

Figure 2.10 Wide Area Network

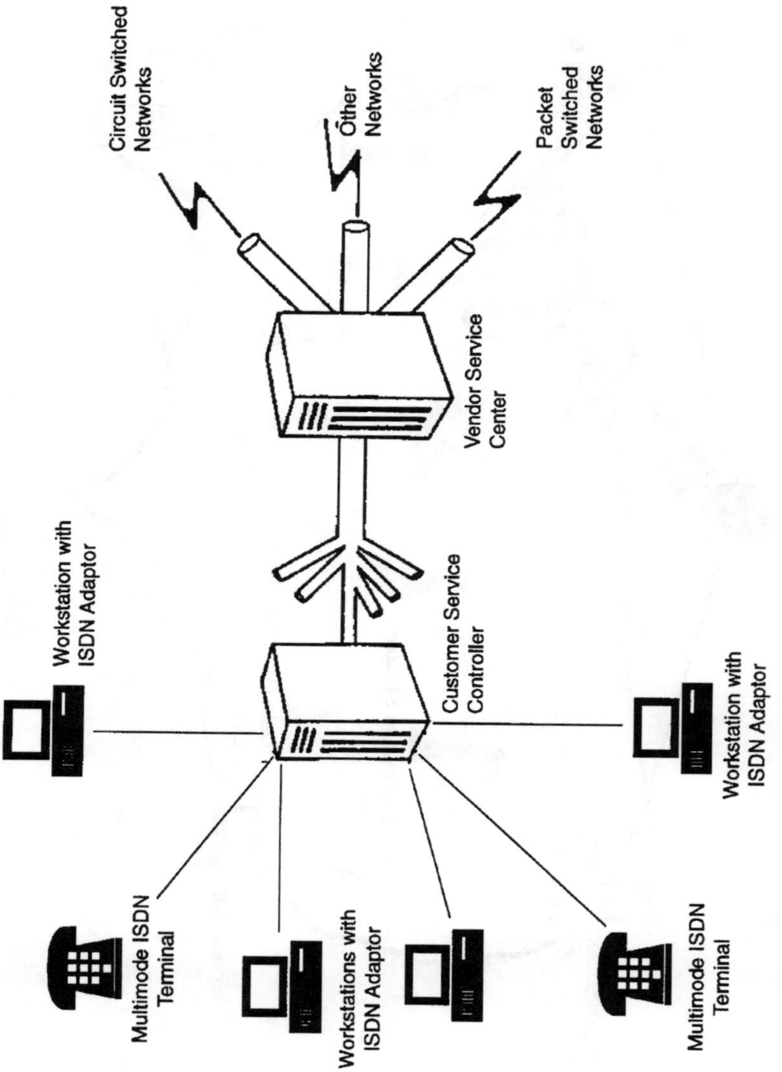

Figure 2.11 ISDN-Based Network

several trials, beta tests, as well as small-scale applications. The high hopes associated with it waned partly because large scale applications had not been developed.

This tenuous technology has spawned many humorous interpretations ranging from "I Still Don't Know" to "It Still Does Nothing." Nonetheless a new outlook on ISDN is surfacing due to the insatiable demand for high-volume information generated by the Internet. The recent upsurge of interest in ISDN is due to three primary factors. First of all, because of the new standards, ISDN can now be provided over the existing copperwire telephone infrastructure. Second, price is getting more affordable. Third, it delivers the high bandwidth needed for the World Wide Web and other network activity.

ISDN embraces the totality of information provision. It evolved from the telephone industry's Integrated Digital Network (IDN), which provides digital connectivity for supporting a range of services that include voice, video, data, telemetry, and facsimile. To fully comprehend the scope of the concept, it is useful to discuss its elements separately. First, it is *integrated* in that the same structure can be used for a multitude of communications modes, including voice, data, and video. Second, it can be used to cater to several *services* that an institution may need to function effectively, for example, data processing services, telephone message communication services, and image or videoprocessing services. In an ISDN environment, all these services use the same transmission medium as well as similar input, storage, and output devices. Third, ISDN transmission is *digital* as opposed to the conventional telephone's analog transmission. In addition, the switching mechanism is digital whereby bits rather than volts are switched. Finally, it can be conceived as a *network* that permits devices to send, receive, and acknowledge messages to and from other devices. ISDN is not entirely new, for the process of creating digital lines and switches that are its foundation has been going on for more than a decade.

ISDN Technology

ISDN is not proprietary to any single vendor. It is based on international principles and standards. Steps to encourage ISDN development started in 1980 when the International Telecommunications Union's (ITU) Consultative Committee on Telegraph and Telephone (CCITT—now called ITU-T) decided to promote standards to accommodate it. Using ITU-T standards, as well as national standards like TCP/IP, devices can transmit and receive data, voice, or image.

Currently ISDN has two levels of interfaces: the low-volume basic rate interface (BRI) and the high-volume primary rate interface (PRI). BRI is composed of a two-wire loop between the vendor's central switch and the user's premises, terminating at an ISDN compatible device. Two 64 Kbps "B" (bearer) channels and a 16 Kbps "D" (Delta) channel are multiplexed over a single 192 Kbps pair of wires. This format is often referred to as "2B+D." It is full duplex. The bearer channel, B, can be used for either voice or data, while the D channel is for signalling (call setup and acknowledgement), but not for voice transmission.

The PRI interface is like the BRI in composition for it has both B and D channels. However, PRI may support 23 B 64 Kbps channels and one 64 kbps D channel (30 B and two D channels in Europe) multiplexed over a 1.544 Mbps digital interface. In the U.S. it is known as the 23B+D. PRIs are normally connected to Computer Branch Exchanges (CBXs) or Private Branch Exchanges (PBXs). PBXs and CBXs connect organizational telecommunications systems to the telephone company's (carrier) distribution center called the "Central Office." The essential difference between the BRI and the PRI is the volume of throughput.

The equipment needed for ISDN installation and operation depends on the functions it is expected to support in an organization. Some of the equipment currently marketed include: terminal adapters, phones, facsimile machine network interface cards, and audio-video conferencing devices. For these to operate, ISDN protocol software is required to interface the local network with the public switched telecommunications network. A basic installation may be traced from the telephone company's digital switch at the Central Office (CO). Signal transmission is accomplished by a pair of copper twisted wires connected through a U interface in a bidirectional mode. Some devices may connect directly to the U interface. A S/T interface may also be used (see Glossary). S/T is a four-wire multipoint through which many devices may share one ISDN circuit. Data between the U and the the S/T interfaces is converted by NT-1 network terminator. A terminal adapter is the final requirement for the end-user computer equipment (PC or workstation). In most cases, the end-user is responsible for providing the NT-1, S/T interface, and terminal adapter. The rest is provided by the telecommunications company.

ISDN Relevance

Since it combines delivery of voice, data, and images, ISDN is ideal for transmitting full text as defined in this text. McDonald's Corporation (the fast food chain) made it a corporate standard for information transfer from mid-1988. Other organizations that have adopted it successfully include the Lawrence Livermore National Laboratory in California and the Baltimore School District in Maryland. Libraries and information centers can use ISDN for accessing image databases, as well as fulfilling interlibrary loan requests. Document delivery services currently using fax or courier services should investigate using ISDN as a faster alternative. Both the sender and the receiver must have ISDN connections using appropriate devices for an effective transmission to occur.

ISDN's Implications for Information Professionals

Information professionals must collaborate in devising applications that utilize ISDN for resource sharing. Obvious areas include the library, document delivery services, computing and information systems, and telecommunications. As the parent organizations adopt the ISDN for computer aided design (CAD) and computer aided manufacturing (CAM), the onus is on information professionals to suggest ways of utilizing this fast, high-volume information delivery technology for other full text dominated activities. It is unlikely that professionals in other disciplines, like engineering design, will be concerned with information processing in general. There are other emergent technologies being tested for full text information delivery such as frame relay; asynchronous transfer mode; and cable modems. Cable modems are perhaps the most prospective for an average organization with modest income.

Like any new technology, cable modems have limitations to full scale implementation. First, cable companies developed in a rather haphazard manner and standards for interconnection have not yet been developed. Second, the networks were developed with one-way communication in mind which was perfected. The reverse communication from the end user is fraught with "noise" which may be picked up from other devices within the building. Third, deployment is not universal and many individuals and organizations will be left out of the loop.

References

Bellamy, John. *Digital Telephony, 2nd ed.* New York: John Wiley, 1991.

Buck, R. G. "ISDN and IN Working Together." *BT Technology Journal* 13, no.4. (April 1995): 113–119.

Clark, Martin P. *Networks and Telecommunications: Design and Operation.* New York: John Wiley, 1991.

Cox, Donald C. "Wireless Network Access for Personal Communications." *IEEE Communications* (December 1992): 96–97.

Frankel, David. "ISDN Reaches the Market." *IEEE Spectrum* 32, no.6. (June 1995): 20–25.

Hudson, Heather E. *Communication Satellites: Their Development and Impact.* New York: The Free Press, 1990.

Hunter, John D. and William W. Ellington. "ISDN: A Customer Perspective." *IEEE Communications Magazine* (August 1992): 12–23.

Kessler, Gary C. *ISDN: Concepts, Facilities, and Services, 2nd Ed.* New York: McGraw-Hill, 1993.

Learn, Larry L. "The Role of Telecommunications in Library Automation: Past, Present and Future." *Library Technology Reports* (July-August 1990): 503–515.

McGovern, Tom. *Data Communications: Concepts and Applications.* Scarborough, Ontario, Canada: Prentice-Hall, 1988.

Srinavasan, S. "A Comparative Study of National ISDN Versus International ISDN Installations and Tariffs." *IEEE Network* 9, no.3. (May–June 1995): 22–26.

Tropiano, Lenny and Dinah McNut. "How to Implement ISDN." *Byte* 20, no.20 (April 1995): 67–68, 70, 72, 74.

U.S. Congress, Office of Technology Assessment. *U.S. Telecommunications Services in European Markets, OTA-TCT-548.* Washington D.C.: U.S. Government Printing Office, 1993.

3. Open Systems, Connectivity, and Information Access in Cyberspace

Open systems underlie all practical efforts to connect disparate computer systems for the purpose of information transfer. Implementation, however, is full of obstacles caused by corporate proprietary considerations, issues associated with intellectual property, and national sovereignty. In spite of such challenges, the level of connectivity achieved on the Internet testifies to the feasibility of national and international computer dialogues. The imperative for universal dialogue and exchange of information has defied national barriers and facilitated connectivity regardless of hardware and software differences among respective networks.

The mid to late 1990s introduced another concept: *interoperability* beyond connectivity. Information on demand necessitates that computer systems not only connect but provide seamless interoperability in order to enable the accessing user to identifiy and retrieve information from a remote site. Thus, efficient full text information delivery depends on open systems and the interoperability of computer systems in dialogue. This chapter analyzes open systems, connectivity, and interoperability of information sytems as implemented by protocols and standards.

OPEN SYSTEMS, CONNECTIVITY, AND INTROPERABILITY

Open Systems

The term *open systems* carries two interrelated connotations. First, to an information systems analyst, an open system implies one that can access and be accessed by other systems with minimal effort. In this context, it is a computer to computer, or network to network, idea. The "openness" of a system implies that it is transparent and accessible to legitimate internal and external users. It is both a theoretical as well as a pragmatic construct. It is theoretical in that within an institution,. internal members will always have access to records which are off limits to outsiders. It is pragmatic in that two collaborative information systems can pass messages to and from each other given appropriate hardware, software, and protocols. This connotation has been marred by a serious inherent conflict as systems have developed. While systems must talk to each other for the benefit of users, vendors would prefer to differentiate their products in order to maintain competitive advantage. The conflict may be resolved when the value of openness is greater than the value of product lock-ins. With value-added packaged offerings encompassing planning, product, service, and support, vendors may still thrive in an open system environment.

Second, open systems have a human factor connotation that requires designing systems which will be accessible to all people. This is part of the U.S. federal government's promotion of open information systems architectures. In 1994, the Committee on Science, Space, and Technology, of the U.S. House of Representatives, issued a report entitled "Realizing the Information Future," submitted by the NRENAISSANCE Committee of the National Research Council which crystallized this connotation. To paraphrase it, open architecture—the foundation of open systems—makes it possible for varied information services to be provided from suppliers of all kinds to customers of all kinds across network service providers of all kinds in a seamless accessible fashion. This context implies universal uninhibited access to information. As discussed in later sections of this chapter, such open systems rely on implicit or explicit agreements among network administrators to exchange information. Protocols and standards codify such agreements. The open systems concept is closely related to connectivity.

Connectivity

The concept of connectivity between disparate computer systems was born in the late 1980s when relatively powerful workstations came to the computer market. Prior to that time, computer systems were dominated by mainframes or minicomputers. A relatively small number of computer manufacturers operated in the market, including IBM, Digital Equipment Corporation, Honeywell, and Hewlett Packard. Time sharing, using dumb terminals, was the mode of access to computer systems. (see figure 3.1). Who-

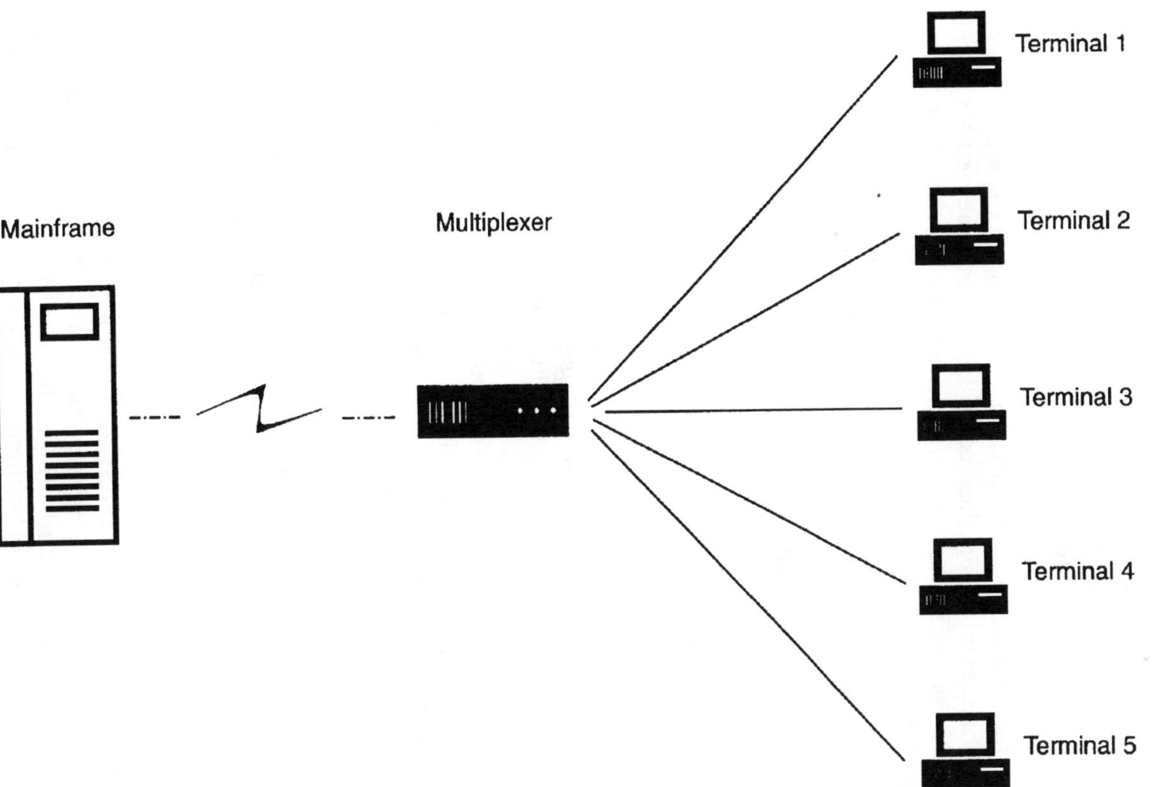

Figure 3.1 Time Sharing

ever wanted to use computer facilities had to go to a communal terminal pool. Logging on to the remote host was done through hard-wired terminals directly connected to the host computer. Terminals were connected to the host through an intermediate multiplexor. As a result each computer system had a unique operating system. Incompatibility was expected between different brands. Some manufacturers had incompatibly among different types of computers. For instance, minicomputers would not share files with mainframes even when they came from the same vendor. Large computer systems also used proprietary access terminals manufactured by the same vendor, or clones. This meant that whatever semblance of full text delivery that existed at the time had very restricted sources that were not readily sharable.

In the late 1980s, the use of modems, gateways, and terminal emulation software made the PC act like the proprietary "proper" terminals for a given system. In addition, various gateways were also used to enable PCs to access larger computers as if they were dumb terminals. The main advantage was that a user could use PC applications like wordprocessing or spreadsheets, or even tap intensive number crunching statistical applications that resided on the mainframe, using the same PC. The mainframe or minicomputer installation still had the advantage over PCs because of faster speeds, larger RAM/ROM memories, and larger storage capacities. In contrast, PCs lacked the requisite speed, memory, and storage for large applications. They were thus not entrusted with critical institutional data. In terms of full text delivery, PCs acting as dumb terminals delivered or received portions of the full text. Much of the large output of full text work was printed rather than displayed on the PC terminal.

In the late 1980s, but particularly in the 1990s, PC workstations began competing in almost all aspects of information processing with the larger computers. Information systems managers touted distributed processing whereby PCs did much of the processing that had previously been reserved for the larger computers. Downward migration of applications became common. Furthermore, the networked PCs on local area networks accentuated the process of off loading applications from a central mainframe-based source to PC-based LANs. The need for connectivity became critical and devices like gateways and bridges flooded the information market. In other words, development of LAN software peaked at the same time gateways for connecting networks became invaluable commodities. These developments laid the foundation for full text information exchange among networks via gateways (the black boxes) which provided language and protocol interpretation. The distinction between large and small computer systems became blurred, and in some instances simply disappeared.

The desire for PCs to access multiple hosts created and bolstered the need for connectivity. PC-based LANs literally sprouted up on the information scene. LAN-to-LAN and LAN-to-host connectivity became vital for sharing information on the mainframe host (see figure 3.2). Connectivity issues were resolved by terminal emulation for host connection. Bridges or gateways between LANs using both networking software and hardware solved the interLAN connectivity problem. Efforts to provide effective internetworking paved the way for full text data transmission among disparate networks. Interoperability requires even closer operational relationships among computer networks.

Interoperability

Interoperability evolved logically from connectivity. It implies that the two or more information systems in dialogue have seamless reciprocal access to each other's files. This concept is both a software and hardware implementation. It is an ideal invaluable for full text information delivery across different computer platforms. In practice, there are no systems that have achieved absolute interoperability other than those from the same vendors and running identical operating systems. In discussing interoperability, we again revert back to the problems and issues associated with user access and vendor product differentiation for market share. Interoperable systems are vital for transparent exchange of information among users, but tend to minimize product identity and may lead to the loss of loyal or captive customers.

Transparent, apparently seamless, interoperability among disparate systems and networks has been implemented through intensive use of conversion software built in gateways, protocols, and standards. The UNIX "tribe" of information systems has had a high potential for achieving interoperability due to the portability of the UNIX operating system. However, UNIX is available in more than fifty flavors which are not "exactly" the same. In other words, interoperability will continue to be seriously impeded by vendor proprietary market considerations. Much of its salvage will come via gateways, standards, and protocols.

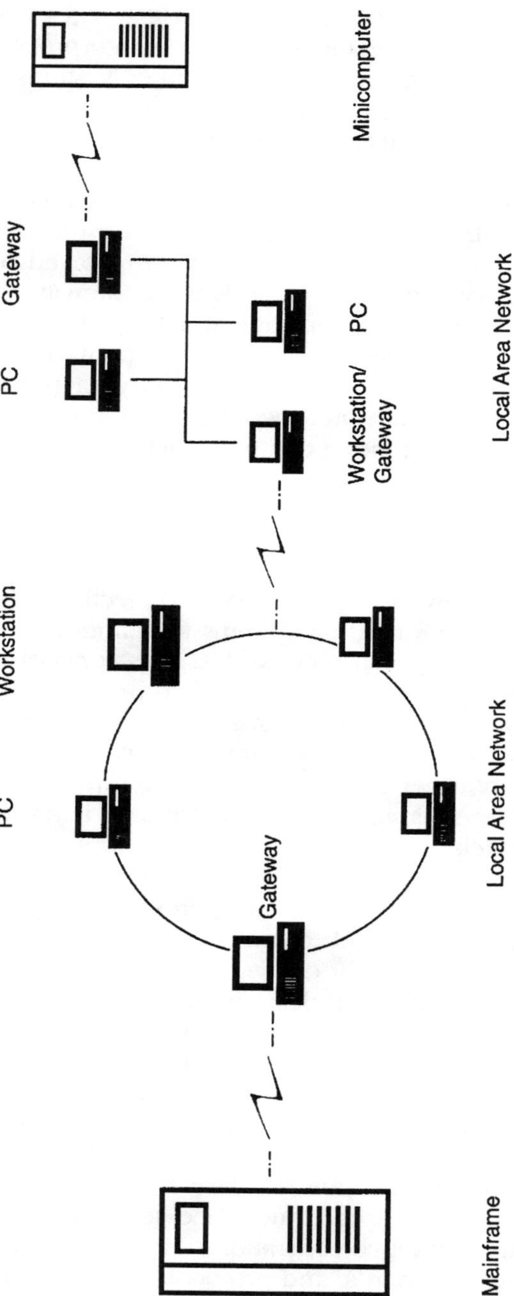

Figure 3.2 Network Connectivity

PROTOCOLS AND STANDARDS

Protocols

While this section does not purport to provide a comprehensive discourse of protocols, it gives the reader adequate information to grasp the rationale for using protocols in information processing and highlights the functions of protocol suites in full text information delivery systems. A telecommunications protocol is a set of rules or conventions which are established for the purpose of computer-to-computer or network-to-network signal exchange. The fundamental goal of protocols is to ensure the accuracy and consistency of the messages exchanged. The term "protocol" has been borrowed from the diplomatic corps where it is used to regulate the way relations between nations are conducted. At national diplomatic ceremonies, for instance, foreign diplomats are advised on how and when to present their credentials to the host head of state. In addition, there are set rules on who responds to the head of state's speech at banquets and other official functions. In diplomatic circles, therefore, protocols provide order and effectiveness when conducting international affairs, so that messages are not misinterpreted.

Telecommunications protocols serve the same function between computer systems. They regulate message transfer to make sure that information systems in dialogue have effective exchanges. When properly implemented, they ensure that what is communicated is what is received by the communicating computer systems. When network protocols are defined, system vendors can design and manufacture products for multi-vendor environments. One of the issues associated with protocols on an international level is the integration of national, international and vendor-initiated protocols. However generated, protocols may be divided into two general groupings: high-level and low-level. Generally, the low-level sets of protocols define how signals are communicated from one system to another, using physical connection media which may include cable, satellite, or microwave. High-level protocols define how applications will be implemented to enable the human operators at both ends to send and receive intelligible messages.

Current national, international, and vendor-generated protocols support a layered format, in which each layer has clearly defined functions that facilitate the modular design of standards for a suite of protocols. Five attributes distinguish layered protocol architectures: transparency of layers; layer packages or "envelopes;" layer independence; synergism among the layers; and reliability for individual layer performance.

First, the transparency of communication characteristics between layers minimizes redundancy from the bottom up. It ensures that what has been activated at a lower level is taken for granted (transparency) and need not be restated at the next level. For example, the layer that deals with file transfer need not specify how the networks physically connect to complete the transfer.

Second, information about the message to be transmitted and how it will be treated at different layers may be envisioned as a package or "envelope." Protocol suite envelopes are used to encapsulate the requisite information from sender to receiver. At the sending site, data about the next layer is included in the datagram envelope which is passed on. The next layer in turn has the data on the following layer, and so on. At the target site, envelops are removed and unwrapped in layer sequence until the uppermost layer software presents the "kernel" message as originally sent.

Third, each layer must be independent of other layers (mainly because of modularity). This means that higher layer protocols may supercede different lower layer architectures. For example, the protocol specifying signal transmission (transport layer) does not affect the way information represented by transmitted signals is displayed at the receiving end (presentation layer).

Fourth, the layers should have a synergistic or holistic interrelationship in spite of the independent function. To effect a transmission, the respective layers which are at play for a given session must be operational for the communication event to occur. Layers thus depend on previous and succeeding layers to perform their part of the communication task for the whole process to succeed.

Finally, there must be implicit reliability within each layer. Functions defined at each layer must be performed accurately, and in a timely fashion, for the whole session to be successful. Ineffective or inaccurate performance will abort the session. This calls into play the final implementation of protocols and standards within the hardware and software used in the session.

These layered protocols have intrinsic limitations which include complexity. This may lead to extra time taken to perform a transaction, making the session more expensive than it would otherwise be. In addition, each layer normally requires extra processing as it goes through the "envelop" procedure, which may affect the response time of the networks. One of the most elaborate layered protocol suites is the ISO/OSI Reference Model.

The ISO/OSI Reference Model of Protocols

The International Organization for Standardization Open Systems Interconnection (ISO/OSI) reference model was developed by the Consultative Committee for International Telegraph and Telephone (CCITT) of the International Telecommunications Union (ITU). The OSI model was meant to facilitate the generation of international networks standards. Since its inception, a number of protocols and standards have been developed based on it. Its main drawback has been competition from the Transmission Control Protocol/Internet Protocol (TCP/IP), which has been adopted by the Internet.

OSI model architecture divides internetworking for information transmission into seven layers (see figure 3.3). It is regarded as a "Reference Model" because it was formulated as a set of guidelines for designing standards and methods for interconnecting computer networks. Layers are explained from bottom up. The first four layers from the bottom are usually designated as lower layers and are concerned mainly with hardware configurations. The top three layers, ordinarily referred to as upper layers, are essentially software implementations.

The Physical Layer is the lowest on the totem pole and defines how a given computer system may be "physically" connected to a transmission medium. At this level, the signals are passed from the computer to a cable, microwave, or any other data transmission media between computer systems. Definitions of voltages, pin assignments, analog to digital conversion, and other chores that make a connection possible are made at this level.

The Data Link Layer follows message transfer between computer systems over the telecommunications channel. It provides data flow control, to make sure that the target terminal does not get overburdened with incoming data. In addition, it provides for error detection and correction to make sure that the data meant to be sent was the data received at the target site. The Institute of Electronic and Electrical Engineers (IEEE) has developed several standards relevant to this layer, for instance 802.5, for a token passing ring network. The Asynchronous Transfer Mode (ATM) protocol which is used for the high-speed transmission of text and

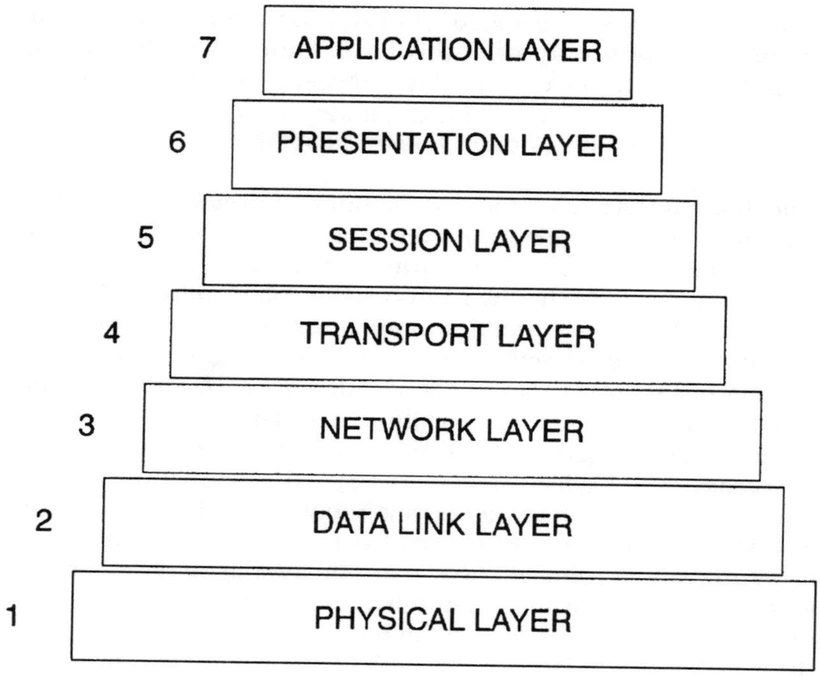

Figure 3.3 ISO/OSI Model

graphics over WANs and LANs is another good example. High-Level Data Link Control (HLDLC) protocol for OSI-based networks is another protocol at this layer.

The Network Layer controls the congestion when data moves from the source node to the destination node. Its functions also include definitions of how data is transferred between networks. The Network Layer performs message packet assembly as they are generated by the sender and disassembly as they get delivered to the target destination. A number of protocols and standards have been developed to conform to the specifications of this level. For example Microsoft's Data Link Control (MSDLC) provides IBM SNA connection to remote printer support. X.25 for packet switched networks is another standard/protocol example. Further implementations are available from other software and hardware manufacturers.

The Transport Layer provides for end-to-end accountability to make sure that a reliable signal transmission environment exists for the networks in dialogue. Transmission error detection and correction are taken care of at this level. Furthermore, protocols and standards developed at this layer detect and adhere to transmission rates between nodes on disparate networks. An OSI example of protocols at this layer is the TP0–TP4 set used for error checking.

The Session Layer provides definitions of how the two communicating network nodes will be conducting the session. It sets the scene for orderly exchange of data. It may define the mode of transmission as alternate two-way—half duplex; or simultaneous two-way—full duplex. It defines intermediate checks that need to be done to ensure the coherent, even flow of messages. It defines parameters or conditions for aborting a running job and restarting a suspended session. Finally, it may assign priority to the jobs being processed. ISO protocol 8327 is an example of a session layer protocol which defines the actual mode(s) in which a dialog between collaborating networks will take place.

The Presentation Layer provides definitions for how the data appears in its final form on the screen, or storage medium, of the target node. It ensures that data delivered is intelligible to the receiver. Data may for example be delivered as ASCII or binary, as defined at this level. Furthermore, data may also be compressed, and protocols that set compression and decompression factors are defined at this level. The layer is concerned with formating the video screen and specifying the audio mode which are very important for multimedia full text information delivery. It defines "virtual terminal" displays, which specifies a preferred terminal at the end node so that the screen is formatted properly according to the terminal emulation. ISO protocol 8823, which defines the mode of presentation preferred at the receiving end, is a good example of one of the relevant protocols at the presentation level.

The Application Layer defines the end-user utilities needed to access the files on the network. Because it deals with end-user whims in terms of requirement, it is very complex. It does not specify what applications are to be done, but it lays down specifications for communications services that should be available so a number of application functions can be implemented. A good example of a relevant protocol is the File Transfer, Access, and Management (FTAM) protocol which specifies communications functions needed for file transfers to be efficient.

In summary, the ISO/OSI model is complex partly because of the nature of the networking community. It is an effective framework within which national and international standardizing agencies, and manufacturers, can develop standards for multi-vendor networks using disparate hardware and software. There are other protocol suites which may be preferred by the general information user public as discussed in subsequent sections of this chapter. Perhaps best known—mainly due the enormous expansion of the Internet—is TCP/IP. Other common protocol suites include: IBM's Systems Network Architecture (SNA), Digital Equipment Corporation's Digital Network Architecture (DNA), Novel Netware, and Apple Computer Corporation's Appletalk. Only TCP/IP will be discussed in detail.

The TCP/IP Protocol Suite

Transmission Control Protocol/Internet Protocol (TCP/IP) is one of the most widely used suites in the world. TCP/IP was developed in the 1960s and 1970s to connect the U.S. Department of Defense contract site computers on corporate and univerity campuses for the purposes of sharing information. The associated network was called the ARPANET. The initial protocol, which connected only three computer sites, was Network Control Protocol (NCP). TCP/IP was derived from NCP and now connects the millions of computers throughout the world that are part of the Internet. It is the most versatile protocol suite currently in existence. It may be used for any operating system, connects any set of computer networks, and uses any of the transmission media on the market. Many of the specific standards that have been developed for the Internet TCP/IP are issued as Requests for Comments (RFCs) and available at the InterNIC, the centralized Internet network information center.

Packet switching is used as the main message or signal flow process mode for TCP/IP. It was developed for the ARPANET to circumvent the circuit switching method which was not fault tolerant. In circuit switching, nodes in dialogue must be connected to send and receive messages to each other. If an intermediate node is used as a relay, messages cannot reach the destination node when the intermediary malfunctions. With packet switching, communicating network nodes need not be directly connected because the information packet can be rerouted via other sets of nodes.

Since it preceded ISO/OSI, many of its concepts and standards are incorporated into ISO/OSI. While it is a layered protocol suite, only four of its layers correspond to the ISO/OSI model. (see figure 3.4).

The protocols are somewhat similar at the lower set of layers of the ISO/OSI reference model. While not formally designated as a specific feature of its structure, the physical layer that specifies the hardware-to-transmission media connection is manifest in some of the protocols working under it. The Serial Line Internet Protocol (SLIP) and Point to Point Protocol (PPP) are used to define network connection over the telephone line.

The Internet Protocol (IP) layer provides definitions for routing messages over sets of networks. It uses addresses combined with packet switching to direct data to a final destination. All other protocols depend on the IP protocol, as it contains the address of the final node or the IP number. This layer also defines the Internet Control Message Protocol (ICMP) that monitors the flow of messages on the network.

The Transmission Control Protocol (TCP) layer is very similar to the ISO/OSI model and determines the speed at which the messages will move from network to network. Initially there was only the TCP protocol which is based on the send/acknowledge prin-

ISO/OSI	TCP/IP
Application	* } Application
Presentation	*
Session	*
Transport	TCP
Network	IP
Data Link	Data Link
Physical	Physical

Figure 3.4 ISO/OSI and TCP/IP Layers

ciple. Using this principle, a sending node waits for acknowledgement of the first package before sending a second package. TCP is rather slow and a faster protocol—User Datagram Protocol (UDP)—was developed as a substitute at this layer. UDP responds only to end-node fault detection. If no transmission fault is detected, an error-free transmission is assumed.

Application Oriented Protocols for TCP/IP. TCP/IP differs from ISO/OSI in defining application protocols. The latter suite uses three layers to define how applications may be communicated on the networks. In contrast, TCP/IP lumps anything above transport as application. Among the protocols that affect full text information delivery are file transfer, telnet, electronic mail, and the hypertext transfer protocol.

Telnet and File Transfer Protocol (FTP) are the most used and the oldest of the pack. Both transfer data from system to system as well as moderate execution of computer programs for remote access. They are provided at the client/user end without regard to the operating system or hardware used.

Another popular application is electronic mail, which is implemented by the Simple Mail Transfer Protocol (SMTP) of the Internet. SMTP is very specific in that messages must be coded in a seven bit, US-ASCII format. The message itself must have headers, which contain information about the origin, and the body of the actual message.

Finally, the hypertext transfer protocol (HTTP) defines how hypertext documents on the World Wide Web are formatted and delivered between nodes. Based on multiprocessing, using windowing, HTTP may request several sources and get them displayed using a web browser like Mosaic or Netscape. The speed at which this application may be implemented depends on the hardware used.

In summary, the TCP/IP suite of protocols provides perhaps the most versatile networking environment ever devised at an international level. Since it offers connectivity for all computer systems at every conceivable platform, it is ideal for full text information delivery. Standards developed for libraries and information centers are taking full advantage of it by using systems such as Ariel for interlibrary lending.

Library and Information Center Standards and Protocols

A number of standards and protocols have been developed or experimented with which deal specifically with library and information center connectivity issues. Since the invention of printing, libraries and information centers have traditionally provided access to full text materials in printed hard copy. Books, journals and newspapers were later joined by non-print materials—videos, cassette tapes, and most recently electronic CD-ROM and online databases. Experiments geared toward generating standards and protocols for electronic full text information delivery in libraries have resulted in a number of applicable protocols. There is a slight overlap with regard to remote access protocols developed for libraires and information centers.

Search and Retrieve (SR) and Z39.50 standards were set up by an American and an international standardizing agency. SR was established by the International Organization for Standardization (ISO) Technical Committee for Documentation (TC46), while Z39.50 was set by the U.S. National Information Standards Organization (NISO). They are almost identical except that Z39.50 allows access and resource control during the session while RS does not. The main objective of these standards is to search databases at remote sites and retrieve records. In doing so, four application phases are defined. First, there is initiation which starts the communication session between nodes. Second, a search request is sent from one node to another. Third, presentation of the search result is defined. Finally, session termination is defined. Z39.50 is used for bibliographic record interchange between the Library of Congress and the three largest American bibliographical utilities, namely Online Computer Library Center (OCLC), Research Libraries Information Network (RLIN), and Western Library Network (WLN). The records exchanged are authority records used to catalog materials.

Interlibrary loan protocols have been developed under ISO/OSI. Two standards were developed: ISO 10160 and ISO 10161. Between them, they define the nature of the service and the mode of interlibrary lending. A loan session may involve a one-to-one connection where a borrower contacts a specific lender or a one-to-many connections where a borrower sends a request to borrow to more than one potential source.

In summary, standards and protocols facilitate the exchange of messages by setting parameters for signal transmission. TCP/IP's implementation on the Internet has made it the main network protocol suite used in internetworking. With some ingenuity, many Z39.50 applications can operate under TCP/IP. However, library automation systems have yet to offer full fledged seamless interoperability, exchanging full text data.

References

Aidarous, Salah and Thomas Plevyak eds. *Telecommunications Network Management Into the 21st Century.* New York: IEEE Press, 1994.

Black, Uyless. *Computer Networks, Protocols, Standards, And Interfaces.* Englewood Cliffs, N.J.: Prentice-Hall, 1987.

———. *Data Link Protocols.* Englewood Cliffs, N.J.: Prentice-Hall, 1993.

Cleveland, Gary. *Research Networks and Libraries: Applications and Issues for A Global Information Network.* London: *Ifla,* 1991.

Corey, James F. "A Grant for Z39.50" *Library Hi Tech* 12 no.1 (1994): 37–47.

Derfler, Frank J., Jr. *PC Magazine Guide to Connectivity, Second Edition.* Emeryville, CA: Ziff-Davis Press, 1992.

Henshall, John and Sandy Shaw. *OSI Explained: End-to-End Computer Communications Standards.* New York: John Wiley, 1988.

Holm, Liv A. "Connectivity and Protocols—The Technical Side: OSI and TCP/IP, FTP, TELNET, SR, ILL, Update." *Ifla Journal* 20, no. 2 (1994): 158–170.

Leon, R. E. "The Internet World: Its Protocols and Mainstream Services". *Online & CD-ROM Review* 18 no.4 (1994): 229–239.

Network Protocols, International Conference on: Proceedings. San Francisco, CA: IEEE Computer Society Press, 1993.

Sheldon, Tom ed. *Lan Times Guide to Interoperability.* New York: Osborne McGraw-Hill, 1994.

Stallings, William. *Computer Communications: Architectures, Protocols, and Standards.* Los Alamitos, CA: IEEE Computer Society Press, 1992.

Summerhill, Craig A. "Connectivity and Navigation: An Overview of The Inter-Networked Information Infrastructure." *Ifla Journal* 20 no.2 (1994): 147–157.

U.S. Congress. Committee on Science, Space, and Technology. Subcommittee on Technology, Environment, and Aviation. 103 Congress, Second Session. *Electronic Commerce and Interoperability in the National Information Infrastructure.* Washington, D.C.: Government Printing Office, 1994.

Wallestein, Gerd. *Setting Global Telecommunications Standards.* Norwood, MA: Artech House, 1990 .

4. National and Regional Computer Networks in the Information Chain

The backbone of electronic full text information delivery is provided by local, national, and regional computer networks within the United States. This is also true for most of the industrialized world including Western Europe, Japan and the Pacific Rim. By storing, forwarding and relaying information, networks form a transparent web that enables the information professional to provide clients information on demand. An alternative scenario has the end-user (the client) access remote data banks directly, with no intermediation. As discussed in earlier chapters, this is facilitated by a combination of powerful personal workstations and the telecommunications infrastructure.

There are several definitions and connotations of networks for computer and general electronic communications systems. Perceptions and terminology somewhat differ when networks are discussed from the traditional telecommunications professional's point of view or from a computer science professional's point of view. To the telecommunications professional, the communications channel and its switching devices, such as the private branch exchanges (PBXs), is the center of attention, while for the computer scientist the center stage shifts to the computer and its peripherals. Both of these approaches, however, view networks as complex structures with many components.

Simply defined, a computer network is a collection of computers and peripherals which are interconnected in order to share resources (see figure 4.1). The links between networks are provided by local or national telecommunications (mainly telephone) carriers. This chapter analyzes the generic computer network, emphasizing its role in full text information delivery. Networks should be conceptualized as interlinked electronic structures in the information chain.

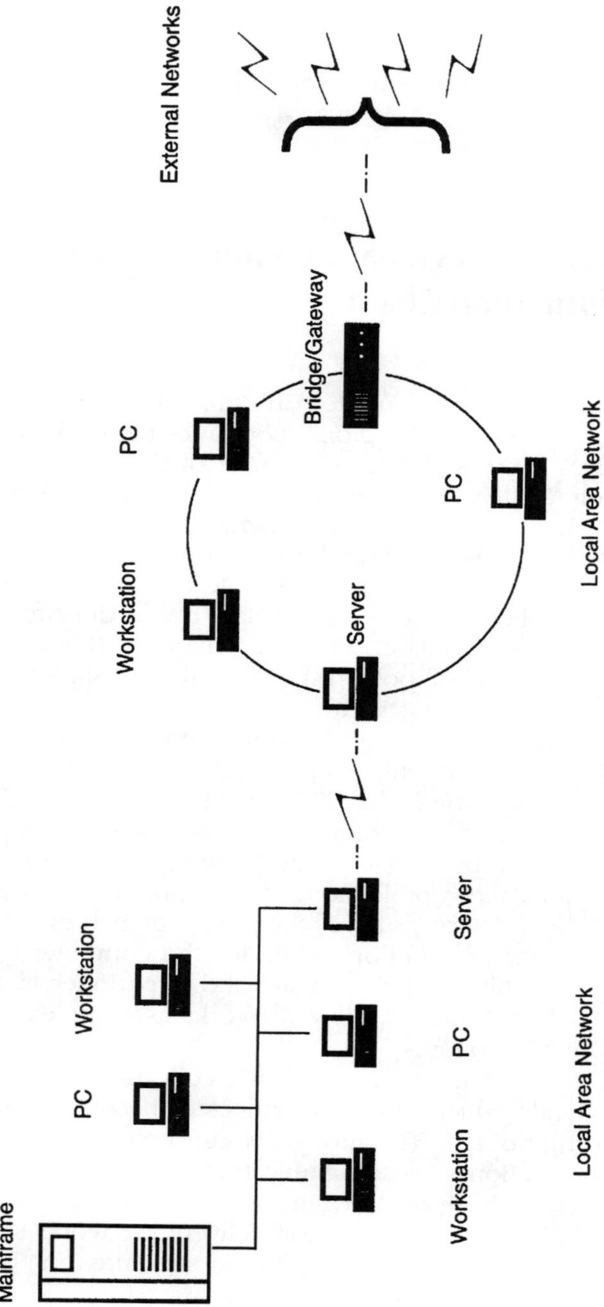

Figure 4.1 Computer Network

RATIONALE FOR COMPUTER NETWORKING

In most working environments, a stand-alone workstation denies the worker the flexibility to access vital institutional and external full text data. However, a networked computer provides an environment for optimizing the information available to the user. Relevant information is identified, accessed, and acquired from information networks on a timely basis, as need arises. Networks provide structures for resource sharing in the widest sense of the concept of sharing. Institutions network their computers for several reasons:

Networks facilitate software sharing. Multiple users connected to a network use one version of a software package which is normally loaded on the network server for public use. The server may have an operating system like DOS or Windows NT, Novell's Netware or a application like WordPerfect. Using the Internet at national and international levels, public domain software is freely available and may be retrieved using the Internet's File Transfer Protocol (FTP) and other tools. The software retrieved may be full text, containing flow diagrams or charts indicating programming logic. With appropriate hardware and software, the end-user can easily download the software to a workstation, with all the accompanying graphics. Computer hardware and software manufacturing companies electronically transmit to their clients segments of software which relate to systems already installed. Such software has been used to patch or update existing software currently running on the user's workstation. Self-contained software applications can be needed for a specific function. Internet utility programs for running Gopher, Mosaic, Telnet, FTP, and e-mail are distributed worldwide in this fashion.

Hardware is shared on networks explicitly and implicitly. Research scientists within the United States share computing power at remote sites and get results transmitted to their own local sites. Such results are usually full text with graphics, and in some cases, sound. International sites have also agreed to act as servers, effect permitting international information seekers usage of their hardware storage and retrieval capabilities via gophers such as Archie and Veronica.

Archie is a combination of resource discovery tools that provide electronic directory service for locating information on the Internet. It was developed by the UNIX Support Group at McGill University in Canada. Archie servers (sites) are organizations that have volunteered to archive information on the Internet and where to find it. Examples are McGill University (archie.mcgill.ca) and Rutgers University in New Jersey (archie.rutgers.edu). Archie sites point out sources contained in "anonymous FTP sites"—where a user can connect without an institutional ID, thus the use of the term anonymous.

Veronica (very easy rodent-oriented net-wide index to computerized archives) was developed at the University of Nevada, Reno by Fred Barrie and Steve Foster. It is a database of gopher menu items gleaned from thousands of gopher servers on the Internet. The simplest Veronica search is a single key word or phrase, for instance, "Elephants" or "Tower of London." It may search all the gopher sites in the world—"gopherspace," or directories that lead to subsequent menus. The main entry point for searching Veronica is the University of Nevada, Reno, but most sites which offer gopher services have access to Veronica.

Furthermore, servers, bridges, and routers (gateways) are implicitly shared by users exchanging full text information on networks. To many information specialists, Archie and Veronica demonstrate international cooperation *par excellence* in the information arena. The information they help to retrieve may be citations, abstracts, tables, or full text.

Data stored in public directories—or bulletin boards—is shared on networks. Again, the Internet has facilitated this type of sharing on an international level. Anonymous FTP sites, which are computer sites that have publicly available files, have allowed their data files to be viewed, and copied according to need. Similarly, electronic journals, which are available in increasing numbers on networks, are examples of full text data documents which are distributed among users. One of the most significant features of these full text information sources is the limited copyright associated with them. In many cases, copyright appears to be completely waived by tacit consent. Another common form of data sharing facilitated by networks is the multiple authoring of documents, whereby authors work together from different computer sites. It is also common to electronically post conference or meeting registration forms on bulletin boards. Interested people download the form, complete it, and e-mail it to conference or meeting organizers.

Networks allow users to share common network utilities and have results of operations either sent to them by regular (snail mail) or e-mail. Shared network printers may accept printing jobs that have been processed at remote sites and print them in various formats with graphics and color. Plotters may do similar functions when complex graphics, downloaded from remote sites, are locally reproduced in black and white or color, depending on the sophistication of the plotters. In addition, networks facilitate sharing storage space on communal disks provided at institutional computer centers.

Computer networks provide flexible schedules for workers whose tasks depend on the computer. Employees can now work at home and submit their contribution to their employers electronically. Networks have thus enabled numerous employees to telecommute and avoid the stress of traffic jams and rush hour. Furthermore, traveling salespeople and other professionals can now verify information from their institutional computer network using laptops in their hotel rooms. They may also submit complex full text field reports with graphics for management attention and higher level decision making. Depending on the sophistication of the laptop, critical business full text documents may be processed in this manner.

ANATOMY OF A COMPUTER NETWORK

The anatomy or the components of a network, and how they interact with each other, is dictated by the network operating system (NOS). The two operating system modes on the market are dedicated-server and peer-to-peer. With a dedicated-server operating system, one or more computers on the network do most of the significant functions of storing network files and allocating network resources to the network workstations. Security, user authentication, and administrative chores are provided by the network administrator. Novell Netware, using SPX/IPX communications protocols, is an example of a dedicated-server mode operating system. On the other hand, with a peer-to-peer mode, the computers on the network have equal status and may allow colleagues to share resources on their computers as well as accessing other users' computer resources. Unix running TCP/IP protocol is in peer-to-peer mode.

A computer network, like any other computer system, has software and hardware components. Software elements include the network operating system (such as Novell's Netware or Microsoft's Windows NT), network interface card drivers, and communications protocols. This software is needed at both the workstation and server ends of the network. The hardware includes the servers, the network interface cards (which are placed in the server and the workstation), and the cable that connects the server to the workstation. A given institutional network may have a server servicing two or more workstations or several servers for multiple workstations and other devices.

Network servers provide a variety of services to the network community. Depending on the mode or environment, they can control much of the network traffic. As a file server, it provides the operating system for diskless terminals and the master application programs so that all users on the network will access the same version of the program. Other activities may include acting as a communications server, thereby controlling access to external networks, print and facsimile services, as well as back-up and archiving services.

User workstations (nodes) are connected to the network using a network interface card. Each node must have software to direct the user service request locally to servers at remote sites. For most network users, the intelligent workstation is the most prevalent access device. Due to increased functionality at the workstation node level, many nodes are usually assigned duties or functions that used to be done centrally, thus resulting in the distributed processing concept.

Connectivity with either local or remote servers is provided by protocols as discussed in Chapter Three. In some cases more than one protocol may be utilized by a workstation to access other nodes on the network. Network driver software programs are installed on workstations to facilitate running one or more communications protocols. Depending on the distribution system, an appropriate protocol is invoked by the driver, loaded into the workstation memory, and a session is initiated. For instance, if a TCP/IP protocol is the mode of communication needed to reach a remote site, it is triggered. Similarly, if a Netware Internet Packet Exchange (IPX) protocol is desired, it is loaded into the memory of the workstation in play. Common network driver standards include Microsoft's Network Driver Interface Specification (NDIS) and the Novell's Open Data Link Interface (ODI). Using either NDIS or ODI, a workstation can load multiple protocols into memory to be used interchangeably as need arises.

Network Interface Cards (NIC) are used as adapters which provide the physical connectivity of the workstation to the network medium used. Specific types of networks require particular types of NICs. For instance, Fiber Optic Data Interchange (FDDI), Ethernet, Token Ring, or the proprietary DECnet would require different kinds of NICs. Cable connection is made through NICs. For example, twisted pair copper wire, coaxial cable, or fiber optic cables join workstations to networks via the NICs. For networks using microwave or other radio media, an antenna is used and attached to the NIC.

The driving software force behind all networks is the network operating system (NOS). NOS enables workstations to share resources on the network. The resources may be broad based—including files, peripherals, and human experts. Network administrative chores like user I.D. verification, server space allocation, and use statistics may be part of the NOS. Operations differ between networks configured as dedicated server or peer-to-peer. With a dedicated server, the programs, files and most functions are provided by the server or a set of servers. Novell Netware is a dedicated server type operating system. With peer-to-peer mode, all nodes on the network have a copy of an operating system which facilitates network resource sharing using a selected protocol(s) for signal exchange. Peer-to-peer examples include: Microsoft Windows NT, Novell Personal Netware, and Artisoft LANtastic.

NETWORK ARCHITECTURE

Network architecture is a broad-based concept composed of three elements: physical layout, methods of cable access, and communications protocols. Protocols were discussed at length in Chapter Three. This section covers the topology and access methods used by network nodes to receive and send messages on a network.

Topology

Topology refers to the network configuration, or the way it is physically laid out, involving placement of nodes and how they are connected by cabling. The term topology was borrowed from geometry where it describes the form of a physical entity. Three design considerations come into play when selecting a topology for a given network. Reliability is of prime concern and different types of topologies offer various levels of reliability. Secondly, routing of messages on the network is usually affected by the topology selected. Some network topologies are rigid with regard to re-routing of messages in the event of an intermediary node or cable malfunction. Third, topology affects response time. Depending on the user needs for which the network is designed, slower response time may be a good trade for greater reliability. There are five common types of network topology.

First, there is the *ring network* (see figure 4.2). In this type of networking, all nodes are strung on the network in a circular mode, hence the name "ring." For this topology, messages are normally transmitted in one direction, with the transmitting node sending information through the intermediate nodes until the target node's address is reached. The intermediate nodes relay the message from node to node. One of the problems of implementing ring networks is that the whole network may be disabled if one of the nodes malfunctions or if there is a break in the transmission chan-

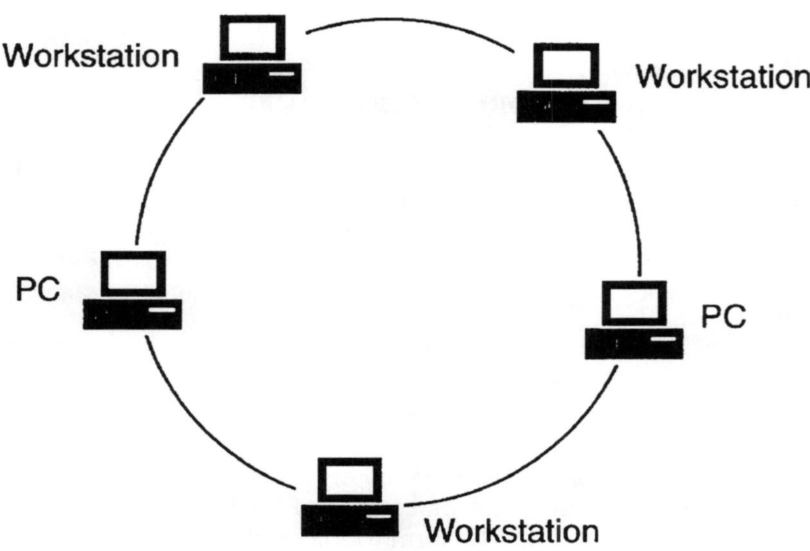

Figure 4.2 Ring Network

nel. Network designers have overcome this problem by incorporating redundancies such as extra backup channels and switching devices to act as relays to replace failing nodes. In case of node or channel failure, the back-up is automatically triggered. While ring networks are currently less common than the other topology types, the IBM and other token rings are logical systems whereby the addressing and transmission follow the ring programming logic even if the nodes are physically not in a ring format.

The *bus network* (see figure 4.3) is also variously called the linear or horizontal topology. It is a popular and relatively simple network to implement. It is a broadcast-based message communication system which permits all nodes to receive every transmission, responding only to those messages which are addressed to them. Since one channel is used by all nodes, the network is stalled when a channel malfunctions. Network designers have overcome this problem by providing redundant channels and bypass switches around faulty nodes. Message security is difficult to manage in a broadcast environment due to the fact that all nodes receive all messages transmitted. Whether done for curiosity's sake or malice, snooping is difficult to eliminate.

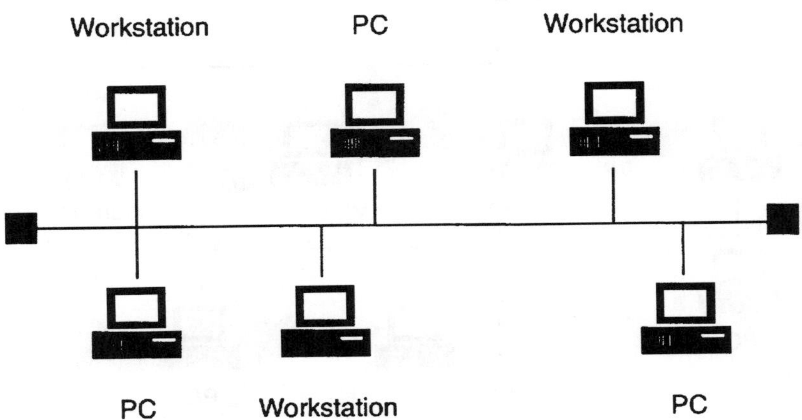

Figure 4.3 Bus Network

The *tree network topology* (see figure 4.4) is sometimes called the hierarchy or vertical network because it is usually designed with a master node which controls much of the network traffic. Some designers provide a distributed control alternative whereby a sub-master node controls those below it. The main advantage of this network is the ease of control and monitoring of network activity. However, reliability may be affected by failure of the main master and sub-master nodes that control the traffic among the clusters. A back-up computer may replace the failed node and a switch may also be used as an override.

The *mesh topology* (see figure 4.5) is relatively new on the market compared to the other topologies. Its main advantage is the inherent immunity to stalling caused by either failed channels or nodes. In either case, an alternative path for re-routing is already built in the basic design. Alternative routes may also be used in case of busy channels or intermediate nodes. The immunity factor makes it one of the most reliable network topologies on the market. One of its disadvantages is the cost, due to built-in redundancies. In addition, it uses complex logic and more than one set of protocols may be needed.

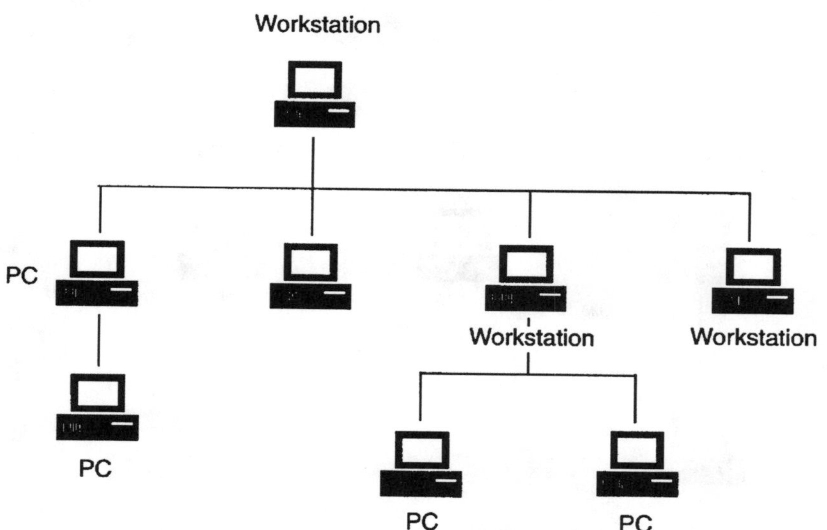

Figure 4.4 Tree Network

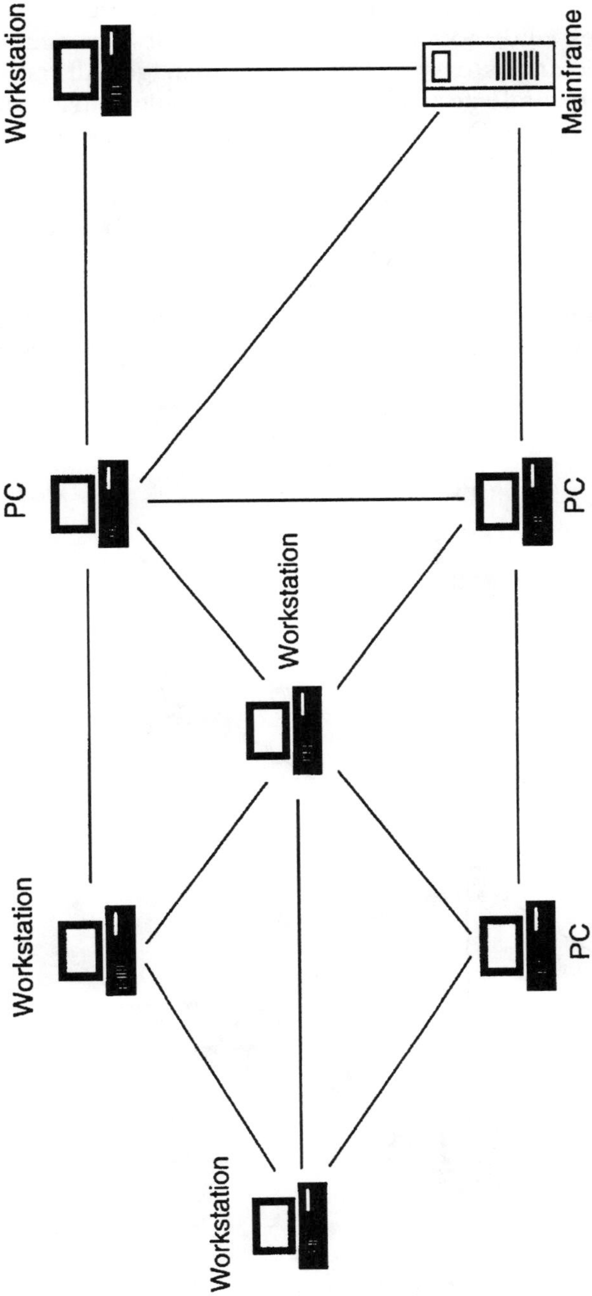

Figure 4.5 Mesh Topology

In the history of networking, the *star topology* (see figure 4.6) is one of the oldest and most common networks. It uses simple programming logic and is easy to design and manage. All traffic originates from a central site that controls and monitors network activity and performance. Comparable to the tree topology, a master node is used with the rest of the nodes subordinate to it. In contrast to the tree topology, there may not be sub-masters and the distributed processing at the individual nodes may be limited. The star topology is ideal for fault diagnosis and management. On the other hand, since it uses the master node concept, when the master computer is down, the whole network is disabled. Most organizations use mainframes at the central node, usually more than two.

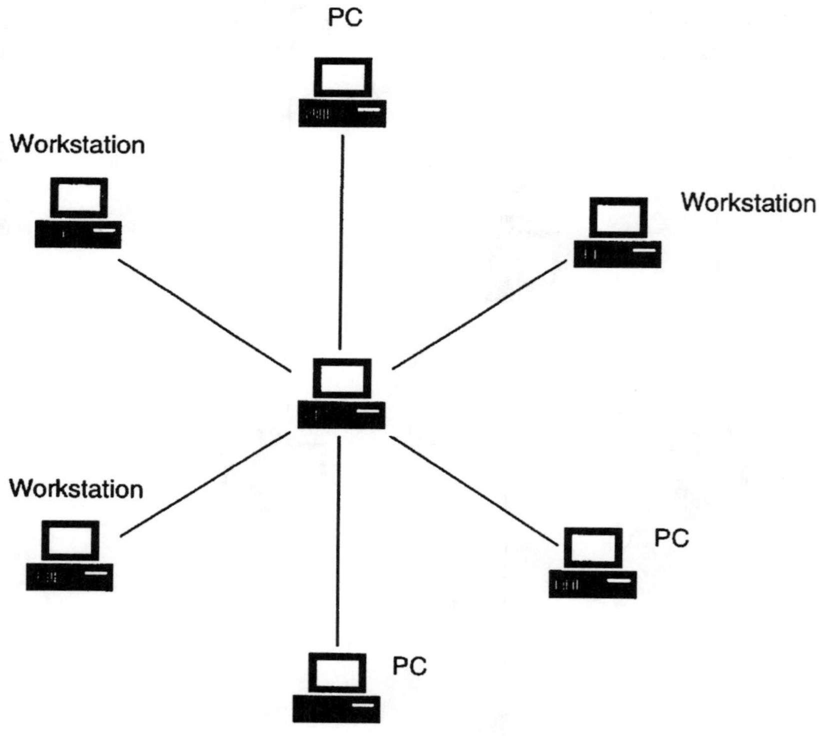

Figure 4.6 Star Network

Network Access Methods

Network access methods are ways of accessing the communication medium for the purposes of sending messages from one node to another. In systems with a master node, like the star network, polling is the main method of initiating a session. The master node or its communications controller polls or queries other nodes to ascertain whether they have messages to send. When a positive acknowledgment is received by the communications controller, a session is set up, a destination node is determined and the packet, or datagram, is transmitted to the target node.

For bus or linear networks which are typical of most ethernet networks, *carrier sense multiple access with collision detection* (CSMA/CD) is used. In this method, a node (workstation) listens (senses) for traffic, and if another node is attempting to send at the same time, it backs off. If two nodes attempt to send at the same time there is a collision and both back off, wait for some time, and send again. Otherwise, if all is well, a node may send its message in a broadcast form, whereby all the stations hear it but the target node is the only one that extracts the message for client use. High traffic slows down response time and thus general performance likewise gets degraded.

Token passing is used for ring networks. Nodes do not initiate transmission unless they posses a token which is no more than a string of coded instructions. When a node wants to transmit, it must wait for an accessible token and by so doing avoid simultaneous transmission with another node.

PUTTING NETWORKS TO WORK

The Enterprise Network

Enterprise networks are fairly new to the information arena. These networks generally belong to an institution and aim to provide information services primarily to a specified internal clientele. It is the basis for electronic full text information delivery in large organizations. Enterprise networks are as varied as the organizations they serve. They may be found in corporate, academic, governmental, and non-profit settings. Depending on the variety and complexity of services needed, it may be a local area network (LAN); a metropolitan area network (MAN); or a continental or global wide area network (WAN). It is typically a multivendor, disparate device and platform network. Since it may comprise more than one network, it is intended to be seamless and transparent to the network user. The seamless feature may result from the use of several protocols, gateways, or bridges to facilitate internetworking.

Distributed computing is a common feature in enterprise networks, whereby some of the network nodes contain and process common data and act as servers to the rest of the organization. To the end-user, the data might appear to be from the local site when in fact it has been remotely accessed, retrieved, buffered on the local workstation, and displayed as the need arises. To avoid costly telecommunications costs on WANs or MANs, some designers choose to store sets and subsets of the most commonly requested data at regional or local centers. Such data can be regularly updated from a central management information system site, but also introduces a whole range of data management complexity.

The Role of the Internet in Full Text Information Delivery

With reports, articles, and entire books written about the Internet, this section is no more than a cursory description of this global network. Such brief coverage may not do justice to perhaps the most significant conduit for electronic full text information delivery man has ever created. Its justification for inclusion, however, is based on the fact that the Internet is gradually becoming the primary choice for electronic full text information delivery among information professionals. Its use has made recent conventional attempts at speeding up information delivery obsolete. The reader should be aware that the "role" of the Internet continues to evolve.

The three major services of the Internet—file transfer, e-mail, and remote logon implemented through Telnet—were described in detail in Chapter Three. All three services provide degrees of full text delivery, with FTP coming closest to "complete text." Another protocol which permits information transfer is the hypertext transfer protocol (HTTP). It is used to transmit a different category of full text prepared using the hypertext language. As the channel for transmission of messages, the Internet is ideal due to its speed. The nature of full text transmitted on the Internet depends on the application that is desired and the protocols at the application layer of the TCP/IP suite that are used to implement it.

To facilitate the smooth operation of the Internet within the United States, a centralized network information center called InterNIC was created in 1992 under the aegis of the National Science Foundation. In 1992, three separate commercial companies were granted NSF five year grants/contracts to provide services to the Internet community. The assignments were: AT&T—directory and database services; Networks Solutions, Inc.—registration services; General Atomics—information services. While the role players and the roles will change over the years, the initial period was vital in establishing and sustaining the global network. These services provide information on how to get connected to the Internet; guides to network tools and resources; and information on seminars or conferences.

Information Resource Sharing Networks for Libraries

Libraries have had more formal recognition in the development of the U.S. information infrastructure than some people may admit. First of all the remote logon on the Internet targeted library online public access catalogs (OPACs). Second, the Advisory Council for the development of the National Information Infrastructure (NII) includes a librarian who is the dean of a school of library and information science. Third, several congressional hearings on information services have featured librarians making presentations. Over the last three decades, libraries have established cooperative organizations in an attempt to serve the U.S. information seeking clientele. Although they have been referred to as networks or consortia in much of the literature for librarians, they were not computer-based when initially organized. They are mentioned here to show their potential in a re-engineered national information infrastructure.

None of these organizations fit the definition of a network when they were started in the 1970s and 1980s. Some authors have suggested that using the term "network" to describe them was conceptually misleading. It was a misnomer to regard these organizational contractual arrangements as networks as many of the institutional members were not connected to each other by any of the schemes discussed in this chapter.

There are several regional or state library cooperative consortia throughout the United States. The Southern Library Information Network (SOLINET) and the New England Library and

Information Network (NELINET) are two typical examples. Many others offer services similar to NELINET or SOLINET, with minor modifications. The basis for initial organization and expansion was the sharing of expensive resources and the negotiation of volume discounts. Consortia played a valuable role in introducing automation to libraries. Some have sponsored or encouraged libraries to establish and implement computer networks to run library functions. To survive the vagaries of the information age, they must re-tool in order to provide more enhanced electronic-based information on demand.to the members' clients.

Library resource sharing networks are managed by people with information processing expertise, although some of them are operating in a semi-networked environment. Several possible services should be implemented to support their clients. First, co-operative hiring of automation and network services consultants will continue to be needed. Expertise may be expensive to obtain by some of the small-member libraries on an individual basis. Rotating sharable consultants would be invaluable in setting up and maintaining computer-based networks. Second, preparation of online guides could be undertaken by the library information networks. Third, the Internet is literally deluging researchers with information. It may be argued that in some cases there is a degree of overkill. Library information network staffs have the expertise to sift, surrogate, and reformat information. Schemes for identifying regional interests should be devised, and placed on regional data banks. Fourth, acquiring technology (hardware and software) will continue to be expensive for many member libraries and cooperative acquisitions and procurement must be continued. Volume discounts by vendors will continue to be significant in spite of the decreased cost of computing per byte of storage. Through experience, libraries and information centers which belong to networks have discovered that purchasing hardware or software through cooperative organization is much cheaper than going it alone. Networks are likely to continue to provide such financial relief especially for small organizations. Finally, users of networked information need user-friendly graphical user interfaces to access information in cyberspace. Library information networks should sponsor the development of such online interfaces for the different clients that their member libraries serve.

References

Baker, Richard H. *Network Security: How to Plan for It and Achieve It.* New York: McGraw-Hill, 1995.

Black, Uyless. *Computer Networks: Protocols, Standards and Interfaces.* Englewood Cliffs, NJ: Prentice-Hall, 1987.

———. *Data Networks: Concepts, Theory and Practice.* Englewood Cliffs, NJ: Prentice Hall, 1989.

Heldman, Robert K. *Information Telecommunications: Networks, Products & Services.* New York: McGraw-Hill, 1994.

Kehoe, Brendan P. *Zen and the Art of the Internet.* Englewood Cliffs, NJ: Prentice-Hall, 1994.

Lynch, Clifford. "Networked Information: A Revolution in Progress." In, Brett Sutton and Charles H. Davis. *Networks, Open Access, and Virtual Libraries.* Urbana-Champaign: University of Illinois Press, 1992.

Miflin, Ingrid and Jean Williams. "Online Catalog Maintenance: The Role of Networks, Computers, and Local Institutions." *Information Technology and Libraries* 10 no. 4 (1991): 363–274.

Peters, Paul Evan. "Networked Information and Services: Next Steps on the Road to the Distributed Digital Libraries of the Twenty-first Century." In, Bret Sutton and Charles H. Davis. *Networks, Open Access, and Virtual Libraries: Implications for the Research Library.* Urbana-Champaign: University of Illinois Press, 1992.

Sheldon, Tom ed. *Lan Times Guide to Interoperability.* New York: Osborne McGraw-Hill, 1994.

Wilson, Thomas C. ed. *Impact of Technology on Resource Sharing: Experiment and Maturity.* New York: Haworth Press, 1992.

5. Institutional Networks in Information Access and Consumption

Institutional information networks are structures set up to acquire, generate, organize, and disseminate information in response to the needs of a given organization. Initially started for specific goals and objectives as perceived by the organization's leadership, they are thus by definition proprietary. Full text information is the primary goal of most organizational networks. Institutional networks fall into four categories: the academic network, the corporate (industrial park) network, the government network, and the non-profit network. These categories share common characteristics, the most pervasive being that they are all plagued by perpetual technological changes. Technology both forces change and threatens stability in organizations.

Five factors have interplayed to effect the growth and development of networks in organizations in the 1990s. First, there is the increased power of the desktop PC. This phenomenon was discussed at length in earlier chapters. It suffices to say here that, due to the PC workstation, individuals can now use sophisticated programs and communicate with other users or access remote databases with ease. Second, the nature of information needed at work has changed from predominantly local resources to resources scattered in the organization and, in some cases, resources outside of the organization. For example, in an organization such as investment banking, international sources may be the prime sources during a given trading period. Similarly, modern high-level research tends to use relevant resources wherever they can be found. With Internet links, full-text files in a London database may be as pertinent and accessible as local databases to a New York City researcher. Third, improvement in organizational telecommunications based on the backbone technology has provided effective information conduits for institutional information sharing. Fourth, the need to deliver fast, accurate services and products whose quality depends on available information pervades the educational enterprise, business, government, and other sectors of society. Institutional networks are ideal to deliver such information services in full text mode. Finally, the advent of client/server technology, whereby a remote server communicates to a local workstation (client) using a collaborative program, has facilitated network growth.

This chapter analyzes the institutional network as a purveyor of full text information. The nature, amount, and variety of information handled by a network depends to a marked extent on the institutional functions and culture. A number of organizations—i.e., banking, insurance, academic research, law, and government literally survive on the full text information they process daily. Only two categories, academic and corporate networks, will be treated in detail because their access policies epitomize the two ends of the spectrum. Academic networks have relatively liberal policies, while corporations are more constrained. Other kinds of networks fall somewhere in between these two poles. However initiated, all networks have a central communication structure to which subnetworks, usually local area networks, may be attached.

THE BACKBONE NETWORK

At the institutional level, this central structure is the backbone network or the core technology of the network that permits connectivity by local area networks, or distribution hubs (see figure 5.1). This backbone forms the basic channel and connects individual buildings or sets of buildings in an organization. It has the fastest internode speed within the organizational network and is usually composed of fiber optic or coaxial cabling. Its main function is to carry signals as fast as possible between the nodes which are typically servers on LANs.

For corporate and multi-campus universities, implementing the backbone concept involves using leased telecommunications components of various types. Microwave, satellite, and all digital data communications lines, like the T1–T4 series, may be part of a complex organizational network. In this case, a segmented backbone may be in place with fiber, coax, or twisted pair copper wire as the media used for some of the segments. Interlinks to other networks, which may be LANs or remote sites, are effected through gateways, and bridges, or distribution hubs. A hub is an electronic switch which connects individual workstations to the backbone.

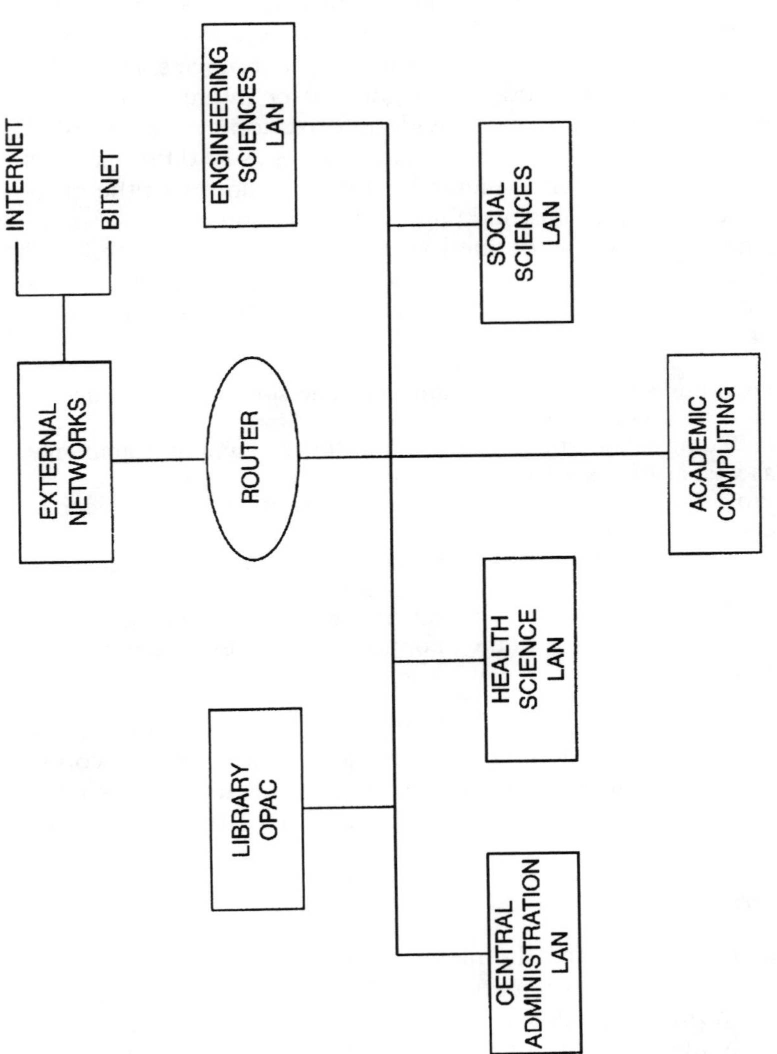

Figure 5.1 Backbone Network

Campus-Wide Information Systems

Campus networks are the boon of college communities as they facilitate access to information never dreamed of before. At the same time, they are the bane of information systems designers as they typically contain disparate information technologies, networks, a variety of platforms, and a variegated user interest base. The literature indicates that campus-wide information systems (CWIS) have been around for at least a decade. Colleges of different sizes and status in the academic hierarchy are grappling with the challenges of establishing, maintaining, and perfecting a CWIS.

In general, a CWIS is defined as a network that is established to provide a seamless, integrated information system that satisfies most user needs on an academic campus. Its basic functions include:

- Providing scholarly information to enable students and faculty to meet their educational obligations.
- Providing administrative information for the campus community as well as the external information seekers.
- Providing local information of a general nature to satisfy academic community interests.
- Providing online remote access to information of relevance to its community from external organizations.
- Facilitating integration of information resources so that one access interface is used for all the information resources on the campus and beyond.

These features are difficult to implement because of the multiplicity of protocols needed on campuses to join disparate networks. The industry has responded by creating innovative devices for internetworking like gateways, bridges, and hubs using protocols like TCP/IP.

Characteristics of an Ideal CWIS

A literature survey and the national study done at the City University of New York, summarized at the end of this chapter, revealed a number of desirable features in a CWIS. They are consistently mentioned in professional journals and are as follows:

A campus-wide information network must be fully responsive to the needs of the clients. What users want and actually need may be somewhat different concepts. User wants can be almost anything, including computer games, all sorts of bulletin boards, and interest groups. On the other hand, needs are tied to the basic functions that an institution performs for society and are thus tied to organizational objectives. If a network administrator is acquiring newsgroups, for example, they must be selected to reflect organizational interest. Deciding whether to have computer games, which ones, and how many, will depend on the organizational culture. Some campuses have some while others do not install them at all.

A CWIS must be designed and implemented in such a way that it optimizes the use of all the resources available at the institution. Resources in this context are broadly defined to include human and material elements. Human resources sharing implies making technology skills available for all people on the campus. A catalog of personnel skills should be established as some individuals who may not be in the conventional technology units may have expertise on computer packages or hardware which they might be persuaded to share. With regard to hardware and software, servers must be optimized in such a way that file, print, and communications servers are equitably distributed within the organization.

Expandability is a must. Servers must have enough processor and storage capacity for current usage and room for future expansion. Given an estimated user expansion rate, they should be expected to last at least three to five years without the need for a hardware upgrade. Channel bandwidth should likewise be on a three to five year estimated capacity projection. If a campus has engineering, medicine, and architecture departments that may need frequent remote access to image databases or a lot of graphics, a fiber optic backbone is mandatory. In addition, the network software selected must accommodate the three to five year expansion projection.

CWIS security must be maintained for users to have confidence in using it for sensitive data. A number of issues come to mind when considering security in a network. First, as an information distribution structure, a CWIS must serve all the university constituencies, but file access must be regulated according to status, whether faculty, student, or staff. For determining access rights, some use the argument that "if it concerns me, I must have access to it, though I may be denied modification rights for it." For example, a student should have access to his or her academic transcript in the viewing mode. This may be relatively easy to implement as a read-only file. Second, security may be considered with regard to general authentication of institutional membership. Some universities subscribe to commercial databases which are subsequently placed on online public access library catalogs (OPACs). While the Internet community may search the general part of the OPAC, subscription databases are normally placed on limited access. Again, this is simple to design using the concept of differential access rights implemented on a menu-style access system. The information processing community has coined the phrase "fire walls" to describe programs or devices which are designed to bar access to sensitive data. Such walls have not been as effective as might be expected when the intrusion is done by experienced systems programmers. In recent years, breaks into institutional computer systems by high school students seems to suggest that absolute security cannot be guaranteed. Perhaps the only watertight security procedure is to have top security information on a separate sub-system that is not accessible to the public, be it a LAN or a time-shared computer.

An ideal CWIS should have a graphical standard interface. Apple and Microsoft have been very successful in using graphical interfaces, thereby simplifying user access to information. In contrast, command driven systems require memorization of instructions in order to exploit the full complement of features that a system offers. Given the choice between a GUI or command-driven system, the average user will prefer the GUI.

A selective design model should be adopted. A number of information systems designers base their design concepts on the philosophy of one size fits all. For simplicity and ease of implementation, they design interfaces that will be used by everybody on the campus. Only the main menus should be the same. For example, interfaces to access medical, engineering, legal, or humanities resources should use some of the professional language and symbols that these various professionals use, especially if the end-user is to access the system with no intermediation.

Sensitivity to the cyberspace environment must underlie the CWIS design and implementation philosophy. Remote access to external information resources will be a major part of the network functions. Consequently, sufficient gateway or bridge capacity must be provided so that users on the campus will successfully navigate the Internet and other networks as the need arises. Similarly, local databases that are meant to be accessed by the public should be appropriately positioned. Conventional wisdom dictates that a public disk is normally provided for this good Samaritan function.

If public access files are to be relocated as is sometimes the case, the opening screen should provide relevant information as to the new locations. Such information minimizes user search time and subsequent frustration.

Both distributed and centralized characteristics should be part of a CWIS. Servers and even user workstations carry many of the network programs and data. For example, the architecture and engineering design images or graphics databank may be located on the divisional sciences server, but it would be accessible to other people at the university who may utilize it. For instance, the School of Fine Arts may want to examine and use architectural drawings. The overall CWIS administrator should be able to monitor and coordinate divisional or campus-wide LAN data. Where college-wide data, such as names and ID files, must be loaded onto divisional servers, regular updates must be made to make sure that all CWIS users entitled to use the data are using the same version. A few universities were lucky to have started with a brand of hardware—the Apple Macintosh—which standardized the campus workstation. However, the average college or university is usually not so fortunate, so there may be more than one platform used on campus. This poses a classic problem of integrating the installed base systems into the CWIS. With the widespread use of the TCP/IP protocol, multi-platform networks pose fewer problems than they used to.

CWIS System Components

Components for a CWIS are similar to the generic network except that these are institutional. The user's workstation (sometimes called the researcher's workstation) is the basic information access device. In academia many different brands of workstations are acquired by various departments due to disparate funding. Before the advent of the campus network, many universities had stand-alone PCs or LANs that were not connected to any other system. They functioned as labs for teaching computer literacy and use of special software relevant to the department's discipline. The situation could be described as the campus information archipelago. In other words, the workstation had a rather limited use. Networking enabled the workstation to become a true work tool for accessing both campus-based and external information.

The high-speed network is the backbone network described earlier and is the main campus communication channel based on one of the common media—coaxial, copper, or fiber optic cables. As mentioned earlier, many institutions are switching to fiber cabling because of its wide bandwidth, relative security, and light weight. This component of the system controls the flow and effectiveness of the campus information traffic. Its bandwidth may dictate the complexity and the nature of full text delivered within an organization. With broadband, a variety of multimedia projects may be implemented.

Database hosts or servers make network services and materials available to clients by sharing software and data. In addition, they regulate "shareware" such as plotters and printers. As the main link between the user and the rest of the world, the server must be in top performance condition as long as the network is open for service. Server down-time must be minimized.

Gateways, bridges, and routers are another feature of the systems components. They have also been discussed in earlier sections, but it is necessary to emphasize their importance in facilitating the connectivity of the disparate campus networks, as well as external networks. Many of them are packet-switching devices, with resident protocol conversion programs. The number and sophistication of these devices may depend on the level of sophistication of the institutional information function.

Human factors are seldom mentioned as system components. Most analysts name hardware and software as the systems components, as if the two can operate without some human intervention. Of course systems can be clocked to start, stop, and do many functions. However a CWIS is developed and maintained to facilitate the information function in organizations. User services must get attention. While it is true that computer literacy is increasing among CWIS users, new systems are developed every day and training users thus becomes a perennial activity. Documentation becomes another important CWIS administrative function. This can be done online or on hardcopy, or both. Some application packages may come with documentation, but in most cases, it must be customized for local system use. Staffing the help desk is another element of user services that is becoming even more important as more non-technical users become regular users of the CWIS.

Another human feature is the technical personnel. Technical expertise should not be the only consideration when selecting staff to be employed on the CWIS. Communication and rapport with the end-user is in some cases more important than what an individual knows about the system. Some information professionals argue that high tech should result in low human intervention, which should make the finance department happy. The high tech, low intervention principle presupposes a perfect, error-free system, which is yet to be designed. In reality, in spite of online help, a friendly voice is usually welcome when the user is stuck. In addition, the utility of online help to the novice user can be quite limited, especially when he or she cannot invoke the help or is unable to comprehend it.

CWIS Design and Implementation

Three units are important in the design and implementation of a CWIS: telecommunications, computing services, and the university library services. In many institutions, telecommunications and computing services are part of one unit. Many institutions consider this collaboration natural and highly desirable. The University of Maryland at Baltimore and Dartmouth College have successfully used this partnership. The library already deals with electronic full text delivery systems via information databanks like Dialog Information Services, LEXIS/NEXIS, and Dow Jones News Retrieval Service, to mention a few.

In terms of division of labor, library personnel provide expertise in selection, acquisition, and maintenance of the electronic reference resources. As regular service providers to faculty and students, college library personnel grapple with user information needs on a daily basis and their experience is invaluable for the programmers designing user interfaces, relational databases, and flat files for client use. Library and information center personnel also help in selecting terms used in menu driven interfaces or GUIs. On the other hand, the computing and telecommunications services unit normally provides systems expertise in the installation, configuration, customization, and maintenance of the computer/telecommunications systems—the physical and logical design aspects of the CWIS. Training and documentation should be developed jointly by the university library system and the computing/telecommunications services center.

The Role of a CWIS in the Reengineering of Higher Education

Reengineering may be defined as the radical and fundamental redesign of an entire business organization, including job descriptions, basic processes, and the structure of the organization's values and beliefs, in order to achieve dramatic unprecedented improvements in the corporate business performance. Consultants and theorists alike have advised the business and industrial management worlds to break, obliterate, and transform organizations for the sake of improving performance. The concept imposes very radical changes that are fueled by technological advances. Proponents of the restructuring and associated draconian internal organizational changes explicitly acknowledge the inevitable role of information technology. Undoubtedly, the process includes massive, speedy, and timely organization and transmission of full text information.

The MIT 1990s Transformation Model, a hierarchy of progressions from status quo to full reengineering, was developed at MIT's Sloan School of Management (see figure 5.2). It shows the six stages an organization goes through to achieve ultimate reengineering. At the bottom of the pyramid is the status quo stage, whereby an organization, either deliberately or due to unavoidable circumstances, maintains itself unaltered.

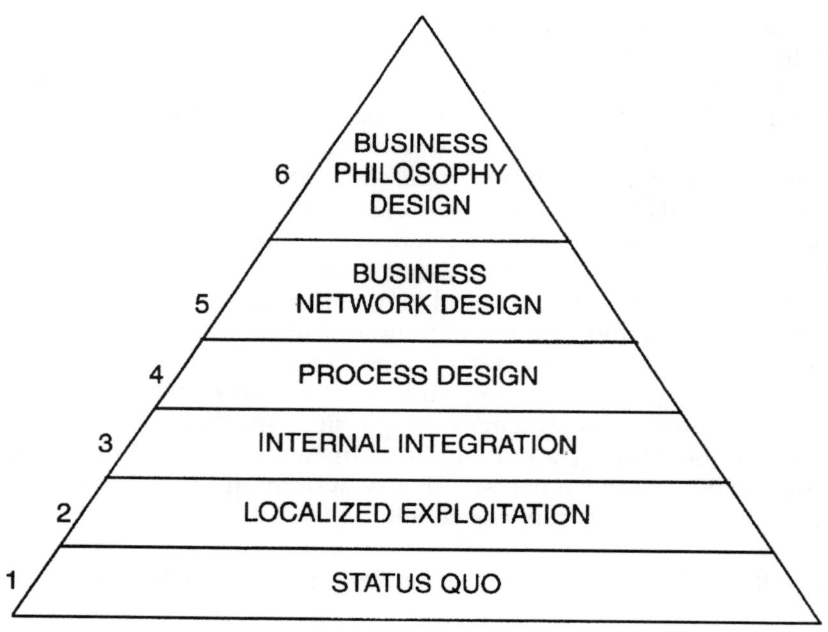

Figure 5.2 A Hierarchy of IT Induced Organizational Transformation

At stage two, movement toward automation is triggered by increasing user demand for quicker, accurate, and timely information. New information systems are introduced to facilitate departmental or cosmetic changes for a given activity.

Stage three integrates formerly isolated systems or even introduces LAN-based subsystems. Machine-readable resources are shared and economies of scale and associated cost reduction are realized. Databases are interlinked to remove redundancies. Computer programs are dispersed via servers, which allows improvement by using the latest and most efficient versions throughout the organization. This stage lays the foundation for full text intra-organizational information exchange. Both individual and unit productivity are enhanced through faster delivery of services and products, thus adding value to items and services generated. Stages one to three are thus evolutionary. The remaining three stages are revolutionary.

Stage four stresses process reevaluation. Consider, for example, the Bursar's process of collecting tuition payments at a college and issuing invoices. Parents or students can check off courses they want to register for, key them in, get a screen display of what they owe, and electronically transmit payments using electronic fund transfer. The Bursar's function would take on a new look with reduction of clerical duties. Similarly, aspects of the central library's interlibrary loan function may be re-evaluated. If a researcher can identify an article needed immediately, he can use the library's remote access system to connect to a document delivery system and have the article electronically sent to his or her workstation. The cost of obtaining the article would be debited to his research grant if applicable.

Stage five considers the whole organization, deciding which processes are essential, which ones can be merged with others, and which should be abandoned. Prioritization is done at this stage, taking into account the organizational goals and objectives.

Stage six is the apex of the effort, culminating when the organization reassesses its philosophy, culture, and role in society. It may divest itself of some of its units, conduct rationalized downsizing, and evaluate its diversification programs. It ultimately redefines itself and its overall *raison d'etre.*

Some have argued that these concepts may not be applicable to higher education, mainly because they will be deemed outlandish and rejected by the board of trustees. Other education administrators have argued the opposite. They have suggested that, faced with competition, rising costs, deteriorating facilities, and reduced government support, the present challenges in education demands new responses and the business process reengineering model must be considered as an option. Admittedly, a complete overhaul cannot be done overnight. However, a number of functions in higher education are amenable to degrees of restructuring using information technology as a catalyst. It is no exaggeration that full text delivery is at the center of all the activities concerned with enabling technology. The Bursar issue, for example, demonstrates what can be done using new technology to reinvent the tuition payment and collection function. In addition, the interlibrary loans function in large academic libraries can also be reinvented.

The development of interactive video and teleconferencing challenge the traditional classroom teaching and contact hours. Distance learning and teleconferencing use similar technology. While teleconferencing is now commonly used in business and industrial transactions, distance learning is new in the education enterprise. One of the current problems of distance learning is cost, but perhaps the most important issues are breaking established practice, protecting jobs, and turf. In addition, such systems have to be tested with regard to delivery of services compared to conventional methods. Do they really add value? Do students learn better? Are faculty more productive?

Finally, electronic full text information delivery systems can provide educators with many options for imparting subject content within the higher education enterprise. Using hypermedia concepts, a biographical history of John F. Kennedy can be delivered in a novel mode. A self-paced presentation can be developed on a computer that incorporates text audio and video. His addresses may appear in an audio window on the screen, and commentaries about him may also be displayed in an audio-visual form. A number of interactive encyclopedias are already using this format of knowledge display. If such a product is provided in the library context, relevant additional readings or reference materials could be provided as supplements. In other words, both processes and content of the higher education enterprise are amenable to reengineering.

THE CORPORATE NETWORK

The 1990s have ushered in tremendous upheavals in how business and commerce are being conducted. The concept of "doing more with less" has pervaded the corporate board room, resulting in the reduction of human resources needed to perform the regular functions in an organization. The key concept has been business reengineering. Enterprise networks designed to deliver full-text information to the worker's desktop have made him or her more responsive to user needs than ever before. Production according to consumer needs becomes the norm rather than production based on schedules set by top management. In addition, total quality management (TQM) becomes the style for conducting business, whereby quality is dictated by consumer demand and decided at the production center.

Many of the issues raised with the CWIS are relevant to the corporate setting. However, the competitive arena makes providing information on demand one of the most important functions in business organizations. Full text delivery thus becomes the prime information commodity eclipsing access to bibliographic citations. Information—whether locally or remotely accessed—must be in usable form for instant decision making.

Reengineering Using the Corporate Information Function

In a modern corporation, the information function includes gathering, processing, and presenting information for decision support. Gathering has traditionally been done by the research department of the corporation, while processing is done by data processing, management information systems (MIS), or information technology, as the case might be. Until recently, internal consumption of information did not get the attention bestowed upon presentations devised for external or public relations use. Information was poorly packaged, using computer paper printouts. Minimal effort was made to summarize, highlight, or use other techniques which emphasize subject content.

The bridge between data processing and presentation has often been lacking, and perhaps the major failure of the MIS shop. This is particularly true of internally generated data. At the middle management level, preparation of the executive summary, using graphics had to await the advent of PC-based software before the executive could do his or her own presentation package. It is now possible to send such full text presentations to customers or field representatives using information technology. For some corporations, need for a field office has been eliminated as executives can now access the corporate database from hotel rooms. Using spreadsheets, a number of MIS activities have been eliminated as the executive or a low-paid assistant may enter the data and generate results at his or her workstation.

With regard to external information, the information center may provide information summaries to the scientist or business executive. Working as an intermediary, the information specialist scans articles and reports and send titles and brief notes to the relevant corporate staff. While this is not done in all information centers, a sizable number of information specialists excerpt or abstract relevant articles to buffer corporate executives from the information glut. This process has been enhanced by a combination of database access, electronic spreadsheets, and modeling software. With these aids, external information is massaged and presented to the executive in usable forms, including statistical summaries and brief reports highlighting the key issues. The network allows the information specialist to proactively acquire, and electronically deliver, information to the client or executive. After browsing the abstracts, some executives may prefer to see the full text which is accordingly delivered. In this case one assumes an enterprise corporate network which has full text delivery functionality. The functions of CWIS and corporate enterprise networks were featured in a recent CUNY study.

Summary of the CUNY Study

With funding from the City University of New York Research Foundation, a study was undertaken from late 1990 to 1993 to investigate the current state of full text electronic information delivery systems in the United States. The impetus to do the study was based mainly on two premises: First, the U.S. information infrastructure was ripe for fully-fledged full text delivery between information providers and users. Second, modern economies put an increasingly higher premium on information as a factor of production. Information professionals need to boost their efforts to provide information on demand via full text delivery systems rather than citations. The main goal of the study was to identify the major players in the arena, and the level of sophistication of existing systems. Its objectives were as follows:

- To conduct a nationwide investigation of the current state of the art.

- To examine technological foundations.
- To identify skills needed for effective operation.
- To assess the current distribution of market share.
- To identify relative costs.
- To identify current practical uses.
- To identify major vendors.
- To assess the information specialists' significance.

Two research instruments were used to gather statistics. In addition, it was intended to reveal information specialists' perception of the significance of these systems in information handling. A different type of questionnaire was mailed out to information delivery systems vendors. This was intended to reveal what systems are actually available or in the pipeline under development. Altogether, 900 instruments were sent out.

To supplement the questionnaires, site visits and interviews were arranged in five regional U.S. cities. The cities were: Chicago (Midwest); Seattle (Northeast); Los Angeles (Southwest); Washington D.C. metropolitan area (East); and Atlanta (South). Cities were selected for ease of access. Individual participating institutions were randomly selected from library and information vendors' directories. Attempts were made to visit and discuss with at least one of each selected type of organizational representative in each of the selected cities as follows:

- Academic libraries,
- State or large public libraries,
- Corporate information centers,
- Bibliographic utilities and service bureaus, and
- Manufacturers and vendors of delivery systems.

Some of the results are tabulated in tables 5.1 and 5.2. As indicated in table 5.1, fax was the most prevalent delivery system, used by 85.5 percent of the sample. A sizable portion of the sample (69.3 percent) were connected to some kind of institutional or campus network. However, with regard to bandwidth signifying the capacity potential of the network, only 23.2 percent had a fiber-optic based backbone as contrasted to 43.9 percent with copper-based networks (see table 5.2).

It was concluded from the research that U.S. electronic full text information delivery infrastructure is in a dynamic state of flux. Information professionals regard the mode of information provision important. As of 1993, fax was the most prevalent mode of

Table 5.1 Electronic Information Delivery Systems

Level of Use	Percentage
Cable Television	17.00
Integrated Services Digital Network	12.00
Optical Storage (CD-ROM)	72.00
Satellite	13.00
Telefacsimile (Fax)	85.50
Teletext	7.70
Videotext	5.50

Table 5.2 Installed Telecommunications Media

Current Campus Network Media	Percentage
Fiber Optics	23.00
Copper Wire	43.90
Hybrid (Copper & Fiber Opt.)	23.60
Other	9.30

delivery according to sample results. Image format, non-searchable text was the most common delivery technique within the information exchange realm. With the development of the Internet as an alternative information delivery mode, the delivery of searchable files is likely to be the norm rather than the exception.

References

Ardoin, Beth and William Weems. "Campus-Wide Information Systems: User Publication of Documents." *Cause/Effect* (Winter 1993): 21–26.

Bernbom, Gerald. "Campus-Wide Information Systems: Managing Information Content." *Cause/Effect* (Winter 1993): 3–5.

Brentrup, Robert. J. Winter 1993. "Building a Campus Information Culture." *Cause/Effect* (Winter 1993): 8–14.

Finnegan, Gregory. "Wiring Information to a College Campus: A Port for Every Pillow." *Online* 14 no.2 (March 1990): 37–40.

Furey, Tim R. et al. "Applying Information Technology to Reengineering." *Planning Review* 21 no.6 (November/December 1993): 22–25, 55.

Hammer, Michael. "Reengineering Work: Don't Automate, Obliterate." *Harvard Business Review.*(July/August 1990): 104–113.

Hammer, Michael and James Champy. *Reengineering the Corporation: A Manifesto for Business Revolution.* New York: Harper Bush, 1992.

Kibirige, Harry M. "Electronic Information Delivery: A Study of the U.S. Infrastructure." In Martha Williams ed. *National Online Meeting.* Medford, NJ: Learned Information, Inc., 1994.

Kovacevic, Antonio and Nicolas Majluf. "Six Stages of IT Strategic Management." *Sloan Management Review.* 34 no.4 (1993): 77–87.

Kriegel, Robert. *If It Ain't Broke, Break It.* New York: Warner Books, 1991.

Lee, Sunro and Richard, Leifer. "A Framework for Linking the Structure of Information Systems With Organizational Requirements for Information Sharing." *Journal of Management Information Systems* 8 no.4 (Spring 1992): 27–44.

Lloyd, Les. *Campus-Wide Information Systems and Networks.* Wesport: Meckler, 1992.

Morton, Michael Scott. *The Corporation of the 1990s: Information Technology and Organizational Transformation.* New York: Oxford University Press, 1991.

Nulty, Peter. "When to Murder Your Mainframe." *Fortune* 128 no.11 (November 1, 1993): 109–120.

Olson, Mark. "We Must Apply the Principles of Business Process Reengineering in Higher Education." *Cause/Effect* (Winter 1993): 54.

Porter, James H. "Business Reengineering in Higher Education: Promise and Reality." *Cause/Effect* (Winter 1993): 48–53.

U.S. Department of Commerce. *National Information Infrastructure: Agenda for Action.* Washington, D.C.: U.S. Department of Commerce, 1993.

6. Transparent Gateway Interfaces in Full Text Information Delivery

The ultimate goal in information seeking, especially for decision support, is obtaining the full text or original substantive document. Decision support systems (DSS) are complex and involve marshaling an array of information sources, sifting through them, and deploying relevant data to key decision makers. While some of the sources may be citations, images, or tables, full text is a basic ingredient of any DSS. Transparent gateway technology is one of the methods used in attaining this goal. The term "gateway" has two connotations. Primarily, a gateway connects two dissimilar networks so that they can exchange signals. It is more system-oriented than user-oriented, although the connection is transparent to the user. Other information professionals equate gateway technology with information search subsystems that facilitate accessing and searching remote systems using a local user interface. In other words, starting at a local access point, a user can "wander" into one or more external sites using virtually the same interface. It is a user oriented connotation, whereby the user's ease of access is stressed leading to an ultimately seamless information universe. This second connotation forms the basis for discourse in this chapter.

Figure 6.1 shows a conceptual presentation of a gateway on a LAN server. Nurtured on hard-copy, index-based information retrieval sources like Business Periodicals Index, Readers's Guide to Periodical Literature, Index Medicus, or Psychological Abstracts, some users are surprised at the availability of an online full text version of an article or a report rather than a citation or abstract. Gateways must aim to provide the original text at the user's workstation, preferably during the same session or soon afterwards.

Information professionals in organizations make a number of assumptions when defining their roles in the information delivery enterprise. First, it is assumed that the language of relevant infor-

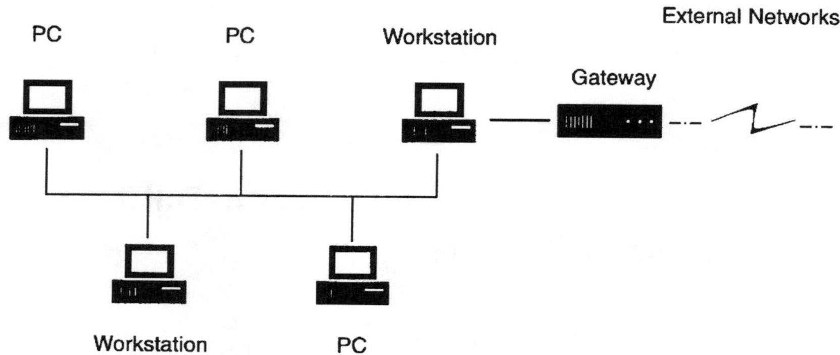

Figure 6.1 LAN with a Gateway

mation sources may be too involved for the client to grasp and conduct effective search and retrieval. Second, the client may be too busy with other organizational functions to devote adequate time to information seeking. Third, technical information is an ingredient that adds value to professionals and may not be an organizational staple. Fourth, there are far too many alternative sources for a researcher to be conversant with intrinsic search techniques. Fifth, an effective optimal search for a complex subject takes more than casual knowledge of sources and an eventual cursory search. Finally, given the training in information access, resource identification, and search strategy development, the information specialist is the best qualified individual in the organization to perform cost effective comprehensive searches. These assumptions leave the end-user little choice but to utilize the services of an information specialist in most information seeking efforts.

Transparent and intelligent gateways attempt to counter some of these assumptions and build a case for substantial end-user independence in information seeking. It is argued that given graphical user interfaces (GUIs), expert systems that help in selecting probable information sources, and user friendly search engines, many of the basic searches can be done by a business executive or researcher with minimal training on how to use a given system. Gateways start with one known information source—say an institutional online public access catalog (OPAC), a CWIS, a corporate information network. From this stepping stone, a user will be directed to another source that may contain the right information and search it with ease. The use of gateway technology-based search systems reduces the level of human intermediation in information access.

This chapter discusses the issues associated with intelligent gateways, routers, and bridges in full text information delivery. Answers are explicitly or implicitly generated in the chapter to respond to standard questions such as:

- Do researchers and corporate executives need intermediaries to find the information they need for decision making?
- If so, what are the levels of searching appropriate for the information hungry seeker?
- Other than emotional considerations, what are some of the justifications for intermediation?
- What are some of the incentives for organizations to use an intermediary given intelligent gateway technology?

THE INFORMATION SEARCH SCENARIO

Information users, even those with some degree of information literacy, are often bewildered by the deluge of potential print and online information sources that they need to consult. The main thrust of our discussion will be the online services. Hundreds of databases or electronic documents are distributed in the U.S. market by commercial vendors like, Knight-Ridder (formerly Dialog), LEXIS/NEXIS, the Dow Jones News Retrieval Service, and the H.W. Wilson Company. Recent information retrieval studies have indicated that no one vendor or database will cover the needs of most individuals, let alone organizations. In other words, to receive comprehensive, relevant, and timely information on demand, more than one source must be searched. Yet, all the available information sources have idiosyncracies in file structures, operating policies, and search languages, which can baffle both an experienced professional and a novice. The same databases, for instance ERIC or Medline, are presented differently by different vendors.

The information needs of users are diverse and can only be satisfied by access to a variety of sources. Given the diminishing public support to information centers and libraries, the increase in end-user searching will help ease the staff shortage in most publicly supported institutions. In addition, downsizing and restructuring will diminish services, including search services in many organizations. End-user training and eventual searching is thus welcome in almost all types of information centers and libraries where such services are provided.

User needs for computer-based information services can be divided into two categories. *First*, there is the need for brief factual information. This requires manual retrieval of information from the so-called ready reference sources, like almanacs, handbooks, dictionaries, census data, and the like. This search can be easily automated. For example, if a sociologist wants to know the population of Minneapolis and the Native American percentage of the city, a simple process is performed to facilitate this type of search on any of the vendors that carry census data files. *Second*, there is the cognitive search that generates comparative statistics gleaned from different sources. This may also provide information used for trend analysis over a period of years. After locating the data, massaging may be needed to present it in a usable decision-support package. Histograms, line graphs, or pie charts may be generated and tailored to the needs of the user. This type of search requires more professional level searching as it combines factual knowledge, professional agility, and a moderate amount of intuition. It is very difficult to automate.

Until relatively recently, most information service vendors offered command driven interfaces to their systems. The norm was searching one system at a time and if a user wanted to start another, he or she would log off and restart again. Vendor software handled communications differently and, in some cases, one had to log off and reboot the computer to start a new vendor session. Two levels of confusion resulted at the search session. First, the user is always taxed to recall log-on commands, and baffled by the "lapsed time" syndrome, where computer systems automatically log one out for trying too many times. Other elements of the quagmire include: user IDs, passwords, and reconfiguring the communications software for half or full duplex. Second, if one succeeds in logging on, one has to decide where to look among all the possible databases in the vendor's collection. After getting to the database, one has to make a selection of terms to use and how to effectively combine them using either AND or OR or NOT, the boolean operators. Both online and hardcopy manuals are available to aid the user. When information is needed for an immediate critical decision, a business executive, physician, researcher, or attorney may not have either the time to refer to the manual nor the patience to check the various alternatives presented. Consequently, if he or she decides to continue unaided, a "hasty" search may be done and with a bit of luck, some relevant material may be generated.

Intermediation is the term used to describe the process of having a middleman between the information source and the end-user. All the requirements for getting information on the system are carried out by the search specialist. While intermediation may be recommended for extensive searches on sophisticated systems, simple searches can be done by the end-user. Some systems like InfoTrac were designed with this mode of searching in mind.

Intelligent gateways minimize end-user sweat and frustration. They eliminate some of the basic log-on routines, and in some cases, provide help in database selection. Commercial software now allows clients to search more than one vendor without logging off the original system. Pro-Search, Scimate, and Easynet are some of these software packages on the market. Pro-Search provides a user-friendly interface to both Dialog and BRS and a relatively seamless transaction for an end-user.

INTELLIGENT GATEWAY TECHNOLOGY

Two basic components make up gateway technology. First, the hardware can be obtained from many computer manufacturers that may have machines built specifically as electronic switching devices or gateways. Common examples include software by Digital Equipment Corporation (DEC), Gandalf, Virtual Microsystems, and Logicraft. Vendor specifications control operational parameters such as how many simultaneous sessions the system can sustain and what access speeds are possible.

Second, the software can be developed using packages that allow multiple access to a database. For instance, Novell's Access Server, Carbon Copy, PCAnywhere, and Everywhere Access allow external access to the databases an institution may choose to provide its users. Actual searching of the database is provided by a search engine (software) which may be bought or leased ready made or custom designed by the system vendor. Another alternative is the use of an expert building program to develop a more sophisticated interface. NEXPERT is used in this fashion. All expert systems are based on artificial intelligence logic and their designers collect knowledge from human experts to figure out what terms, log-on conditions, and database system features to mimic human search strategies. Sophisticated gateway design and implementation on a local basis requires relatively high-level programming skills.

Other considerations that affect the gateway hardware and software performance are normally transparent to the end-user and are defined by the telecommunications environment. They include: the telecommunications channel; transmission media, whether to use fiber optics or coaxial cable or other media; and bandwidth. Some gateways use the client/server technology discussed in earlier chapters. Depending on how the gateway concept is implemented, the quality and sophistication of the access terminals (computers) may affect the speed, and resolution, of the image displayed at the end-user's desk.

User-System Dialog

The actual dialog or interaction between the end-user and the intelligent gateway may depend on how the information systems (IS) or computing services department in the institution view the gateway's role. A number of connotations of the concept are used in the information arena. Within the last decade, terms used to describe the gateway concept have included front-end, interface, post processor, transparent system, and simply gateway. In this chapter, all these terms are included in the generic "gateway" concept. There are various ways of implementing a gateway.

As in the most basic method of providing access to other information systems, the gateway takes you there and leaves you at the mercy of the external system. On the Internet, for instance, Gopher and Telnet are very good examples of basic gateways to other information systems. If you connect to foreign online public access library catalogs (OPACs), you need to know the local language to read instructions and continue.

Another option, sometimes referred to as the front-end, is the sophisticated dialog already covered. It embraces all logon procedures but also includes help to select the database. On the Knight-Ridder system, for instance, the front end involves the use of the Dialindex feature, selection of search terms, and search modification. Dialindex is a Knight-Ridder feature that helps the user in identifying what databases in the vendor's databank have citations on a given subject. Some examples of front ends include Telebase's Easynet and Online Computer Library Center's (OCLC) EPIC system, which is based on Easynet.

Some literature has mentioned post-processing as one of the features of an installed gateway. The connection between the two depends on what the systems designers have decided to put into the gateway by way of functionality. It is legitimate to expect the gateway to incorporate word processing and spreadsheets and other functions to enable the end-user to prepare a usable decision support package. Downloaded material should be easily imported into appropriate software to create customized documents. This feature should be designed into the gateway, since its basic function is to ease the end-user's burden so that his or her effort is directed towards the decision for which information is sought.

Gateway Interface Residence

Where to locate the gateway may not be a local decision. For example, when Dialog and BRS provide The Knowledge Index and BRS After Dark, respectively, they are attempting to provide easy access for the end-user. Their systems have programs at the host which may be reached using standard communications packages.

An alternative is to have the gateway program at the user's workstation. Such an arrangement may give the end-user freedom to customize and be creative in designing the final decision support package. Pro-Search, SciMate, Dialoglink, and The Grateful Med do this. They are designed for simplified user access to the host databases.

Finally, a number of institutions have designed in-house gateways of various complexities. Lawrence Livermore National Laboratory in California has had a gateway in operation for a considerable time. It provides researchers and information specialists access to local and external databases using a standardized interface. Log-on procedures are very simplified for easy entry into a desired database. Many OPACs are being redesigned to incorporate gateway concepts and allow users to access other information systems beyond the local ones.

Selecting an Access Mode

There have been several approaches to gateway interface design. Four kinds of user interaction are commonly used: Command driven, mnemonic images (or icons), alternative options (menus), and user-defined terminology (natural language query).

The command driven mode has been mentioned before in the text. Its main disadvantage remains that the user has to memorize many of the commands to be an effective searcher. Its main advantage is the flexibility it offers for strategy development since boolean operators can be utilized. A search can thus be restructured in several ways to produce optimum results.

Mnemonic images, popularly known as the graphical user interface, use both icons and image correspondence to ease user dependence on memorizing the basic searching terms. Images or icons represent semantic correlations which are easy to remember. Its main advantage remains the ease of use and the memory "prompting" provided by the icons. Its disadvantage remains the use of predefined search parameters which reduce search flexibility. A searcher may not be able to regroup subjects as is done with the command language search technique.

The alternate option choice is characterized by a menu giving selection options. Due to predetermined searcher options, the subjects to search have to be among the options. Relatively new subjects may not be part of the menu selection. At best, a general subject may be selected yielding marginal results.

The fourth design option is user-defined language—commonly called natural language. Many database vendors are now providing natural language as the access mode to ease the complexity of the command language requirement. However, many of the systems are not as precise as either command or menu driven systems due to the broad categories which are bound to be selected by the end-user as search strings.

BASIC CHARACTERISTICS OF A TRANSPARENT GATEWAY

While the concept of a gateway may be implemented in several ways, there are some common characteristics that all gateways should contain. First and foremost is the transparent direct connectivity to external information services. With this function, the system user should never have to deal with log-on procedures common to Knight-Ridder's Dialog, Dow Jones's Information Retrieval Service, or Lexis/Nexis. Even if a sophisticated gateway is not provided, the automatic log-on facility of many of the popular communications packages can eliminate this basic function.

Second, the ability to perform moderate manipulation of the remote hosts files is a desirable characteristic. Given the potential number of relevant sources, a variety of local procedures may limit the number of actual accessible sites. For this feature to be incorporated, local designers must have clearance from the distant sites. Easynet does allow access to twelve hosts using its local query language.

Third, online documentation is offered by most gateways in various forms. Help screens are provided to respond to frequently asked questions. Common queries are concerned with the commands used, particularly semantics, and cost effective combination of the available commands. Database directories and brief search guides like the Dialog Blue Sheets are available from ProSearch. Easynet was one of the earliest gateways to have online human help with its "SOS" command. A live expert staff member gets on the system and responds to user queries. This is the equivalent of the help desk, though the messages are typed rather than spoken.

Fourth, file selection assistance is a desirable feature to include in a gateway. The process involves a clear understanding of the user's query and the available databases accessible to the institution or individual. One of the issues associated with the databases that would make this feature difficult to build into the gateway is the constant changes that databases go through. However, some search aids provided by vendors can be invoked from a gateway and generate an indication of database subject content. Easynet's SmartSCAN; Dialog's Dialindex; and BRS' CROS provide numeric figures that indicate occurrences of the query in the available databases. For example, if a city's public relations officer wants information on "child abuse," a Dialindex search will indicate how many hits are in ERIC, or PSycINFO, which is a psychological abstracts database. With an interactive display, a user can make informed decisions as to what is the most relevant database to concentrate on.

Fifth, attempts have been made to develop gateways that can help with search strategy development. Most investigators concede that it is the most intellectually engaging portion of online searching. It requires knowing the user query in context, grasping system capabilities, and making profiles of alternative matchings. Much of what goes into the work done at this stage is not found in manuals. Most searchers restructure the search after the first set of results, a procedure that would be very difficult to regenerate with a computer. While research in artificial intelligence that generates expert systems has promised delivery of aids in this realm of searching, reliable commercial products have not been marketed to date.

Sixth, information capture is invaluable as a feature to have in a gateway. It is rare for a user to take advantage of all the information as it is scrolled off the screen. Most users download material and print it off for later perusal. Most network packages, such as Procom, Smartcom, or Crosstalk, have downloading and uploading capabilities. With minimal software reconfiguration the information capture feature can easily be built into any gateway.

Seventh, decision support oriented information needs massaging to make it easily usable. If the gateway does not provide word processing and spread sheets or other formatting capability, the data captured should be in a processible form. Conversion subprograms should be in the gateway so that wordprocessors, or any other programs for final processing, do minimal preliminary document formatting.

Finally, an ideal gateway must attempt to provide seamless multihost connectivity. Since its major goal is to simplify information access, switching from a remote site to another site should be easy and hassle free. Some systems, like Dialoglink, allow a user to logon to other vendor systems, although the process may be rugged. Depending on how the contract was drawn, accounting for some of the services, like telecommunications, may be confusing.

IMPLICATIONS FOR INFORMATION PROFESSIONALS

However intelligent, gateways have not reached a stage where they can supplant information specialists. Nonetheless, they are truly helpful to the information professional in that they permit the client, or end-user, to do some basic searches. The information specialist would then use her or his expertise to concentrate on the more complex searches or other functions in the information center or library. He or she would select relevant electronic sources to satisfy institutional needs and also train users how to use them.

The cost of searching by clients who are not information specialists tends to be higher, since they lack the experience to do precise comprehensive searches. In addition, because they are infrequent searchers, they would not do well with complex searches that require restructuring the search strategy after the first set of results. Moreover, when information delivery is incidental to the institution in question as a value adding activity, the staff member may not devote adequate time to information seeking. In terms of opportunity cost, they might be generating more revenue by concentrating on the functions they were hired to do—such as accounting, teaching, public relations, law, or medicine. They would not respond very well to the vagaries and pressures of obtaining information on demand.

Society needs to realize the importance of information as a factor of production. This is true for all types of organizations, including those engaged in the educational enterprise. Productivity is so intertwined with technical information that marginalizing it simply as an overhead does not do it justice. Most of the companies that have successfully re-engineered some of their processes, like Eastman Kodak, IBM Credit, and General Motors, have depended very heavily on information technology as the enabler. Information used in the restructuring endeavor is usually the full text type. Lighting, gas, water, and other utilities that are classed as overhead, are certainly not in the same category by any stretch of imagination.

The issues associated with intellectual property and copyright loom in the future of intelligent gateway technology mounted on the Internet. Site licensing solution whereby an entire corporation's staff are eligible for unlimited use of a copyrighted product as long as they are conducting legitimate organization business, has alleviated part of the problem. Institutional licensing, however, takes on a new turn where the accessing clients are worldwide, with no allegiance to the institution providing an access point. Vendor-to-vendor negotiations that involve bundling, or value addition, to a product before resale have also solved part of the problem, as the resaler handles part of the paper work. Suggestions for a common access and file manipulation language beyond the basic protocols makes information product differentiation difficult. Information processing companies owing their existence to the uniqueness of their product may resist having common access languages, except in strictly controlled conditions.

The Internet and Progress Toward the Year 2020

The meteoric and exponential growth of the Internet in the mid-to-late 1990s somewhat overshadowed gateway technology. Many of the functions of the Internet make it behave like a super gateway. Gopher, FTP, telnet, and even e-mail have the semblance of gateway technology. Transparent access to remote sites is accomplished with relative ease, and full text files are downloaded or uploaded daily. In effect, client information access to remote sites is made effortless and relatively transparent to the user depending on the system one uses.

However, there is an apparent deception with the Internet in that for many of the connections it facilitates, the user is at the mercy of the remote host. The need for seamless connectivity in order to send and receive full text at one's workstation still persists. Many of the systems are in effect incompatible beyond the handshake facilitated by the TCP/IP protocol. The Internet-based gateway will eventually incorporate the Internet, expert systems, client/server, and other technologies to allow the information users to exchange information uninhibited by the constraints of their local system.

As we approach the twenty-first century, full text information is gradually setting operational policy. It defines the parameters for the magnitude and the nature of individual and organizational operations. Accessing it becomes a high premium and gateways that simplify the process are vital to both the information specialist and the end-user. The Internet has in fact accentuated the need for intelligent gateway interfaces as the connectivity it provides sometimes leads to *cul de sacs*. In many cases, the user is left wanting more dialog at the remote site rather than just getting there.

References

Burton, Hilary D. "The Livermore Intelligent Gateway: An Integrated Information Environment." *Information Processing and Management* 25 no.5 (1989): 509–514.

Efthimiadis, Efthimis N. "Online Searching Aids: A Review of Front Ends, Gateways and Other Interfaces." *Journal of Documentation* 46 no.3 (1990):218–262.

Elias, A. W. "Copyright, Licensing Agreements and Gateways." *Online Review* 14 no.4 (1990): 225.

Evans, Ed. "Enabling South Africa: Developing of an Intelligent Gateway." *The Electronic Library* 11 no.6 (1993): 279–282.

Johnson, Jean M. "Asa on the Bell Atlantic Gateway: The Lehigh Experience." In, Martha E. Williams ed. *Proceedings of the National Online Meeting*. Medford, N.J.: Learned Information, Inc., 1990.

Schoenbrun, Cynthia. "Easynet: What Has Become of the Small Giant?" *Online* (January 1993): 52–56.

Sylvia, Margaret. "Building a Gateway for the CD-ROM Network: A Step Toward the Virtual Library With the Virtual Microsystems V-Server." *Microcomputers for Information Management* 10 no.2 (June 1993): 119–129.

Tolly, Kevin. *IBM Lan Gateways*. New York: McGraw-Hill, 1993.

Welsch, Erwin K. and Ellen Shultz. "Developing Gateways to Online Information Utilities." *Library Software Review.* (Mar/Apr 1990): 83–90.

7. U.S. Information Policy Initiatives for Cyberspace

While the concept of a national information policy for cyberspace is relatively new in the information arena, it builds on more than two decades of efforts to provide full text to the information seeker. As we move into an era of full-fledged electronic full text information delivery, some of the ideals that govern access to information may remain the same. For instance, access to usable information on demand by any information seeker remains paramount. However, the medium that enables access, and sometimes the message that is sought, are in a dynamic state of flux. Studies of information policy done prior to the advent of the Internet need to be revised since many of the assumptions and conclusions made about access are now obsolete. Likewise, treatises on information policy must be re-written to reflect technological trends as the rationale for extant policies. For instance, the rationale for depository libraries as a means of guaranteeing access to federally generated information has become questionable if the Government Printing Office (GPO) can establish client/server systems for remote access to its full text documents databases online. Some of the issues and problems about information policy are worldwide although countries tackle them differently.

Industrially advanced countries like Britain, Canada, Germany, Japan, as well as the other Pacific Rim countries in Asia and some less developed countries have engaged in tremendous efforts to develop national information policies in the last two decades. Several countries take the adage "information is power" very seriously. Almost all nations of the world have realized the importance of information as a valuable national resource. It is particularly significant as an economic development enabler, all other things being equal. As a consequence, some regions, for example Europe, have developed regional information processing and dissemination organizations like Euronet/DIANE.

The United States appears to be an exception in that it does not have a co-ordinated "national information policy." This should not be misconstrued to mean that the U.S. polity does not recognize the importance of information in the national economy. To the contrary, some of the most elaborate analysts that have articulated the significance of information in national economies are very highly regarded American intellectuals. Daniel Bell's classic book, *The Coming of the Post-Industrial Society,* was both prophetic and a catalyst to new ways of thinking about the U.S. political economy. In a similar vein, Fritz Machlup and Marc Porat have espoused the concept of the information economy as the successor to the industrial-based economy. These theorists have laid the intellectual foundation for analyzing various aspects of the information arena, including information policy as it has developed in the last two decades. In addition, the U.S. media has for the last 20 years continuously hyped the information age and information society themes.

Thus, it is not due to a paucity of relevant ideas or lack of awareness that the U.S has no readily identifiable national information policy. With regard to legislation, information policy has not been ignored. According to Congressional Research Service, the U.S. Congress passed 339 laws between 1977 and 1990 that are related to information technology. To borrow from John Stuart Mill's conceptual landscape, the U.S. populace appears to value information so much they protect it from central regulation. In other words, in the late twentieth century, information has become so important to the U. S. international competitiveness that it cannot be relegated to the realm of federal regulation. To regulate it would stifle its exponential growth and result in retrogressive economic effects. Another way of looking at information policy development in the U.S. is to invoke Alexis de Tocqueville's classic essay "Democracy in America," in which he analyzed the deliberative way decisions are made in this country. What has changed since the essay is that the loci of power to influence policy has changed from electoral influence to lobbied influence—specifically, corporate interests.

The U.S. National information policy is a mosaic of many initiatives including Presidential directives in the form of Executive Orders, legislation by Congress, Federal Agency regulations, and state provisions. In broad terms, there are three categories of formal governmental initiatives to harness information. First, there is the drive to provide an environment for an effective medium to deliver full text information. This is exemplified by activities that affect the data communications or telecommunications environment. Second, there is the drive to provide regulations and Acts that govern intellectual property. Third, there is the drive to provide an environment for access to government-generated information through the Government Printing Office. For non-government-generated information, there has been the general funding for libraries. The Library of Congress, the National Library of Medicine, the National Library of Agriculture, and the National Archives constitute the most direct form of funding in this respect. In addition, there has been funding for other types of libraries and other information technology-related projects. Most of the activities considered in analyzing information problems or issues have been within this triumvirate philosophical purview.

At another level of analysis, there are five categories of stake holders with differing capacities for influencing policy. They include government, the information using public, information producing organizations, information device producers, and information transport institutions (see figure 7.1). The locus of real power to influence information policy has over the years shifted from the government to the information corporate lobby, represented by members of the Information Industry Association and other business-oriented organizations.

This chapter analyses the highlights of the policy initiatives and provisions developed in the last three decades which have laid the foundation for electronic full text information delivery. The concept of the global village that was developed in earlier chapters is even more pertinent in this chapter, for much of the national information policy must take cognizance of the international arena.

HISTORICAL PERSPECTIVES

In order to cast information policy initiatives in a cyberspace perspective, highlights of legal provisions are categorized into three groups. Those that emphasize the medium, those affecting the message and, finally, the ones impacting access to information. Many of the original laws and regulations that affect information policy are being amended to account for the technological changes. The main problem with the legal revision is that it has not been able to catch up with the technology.

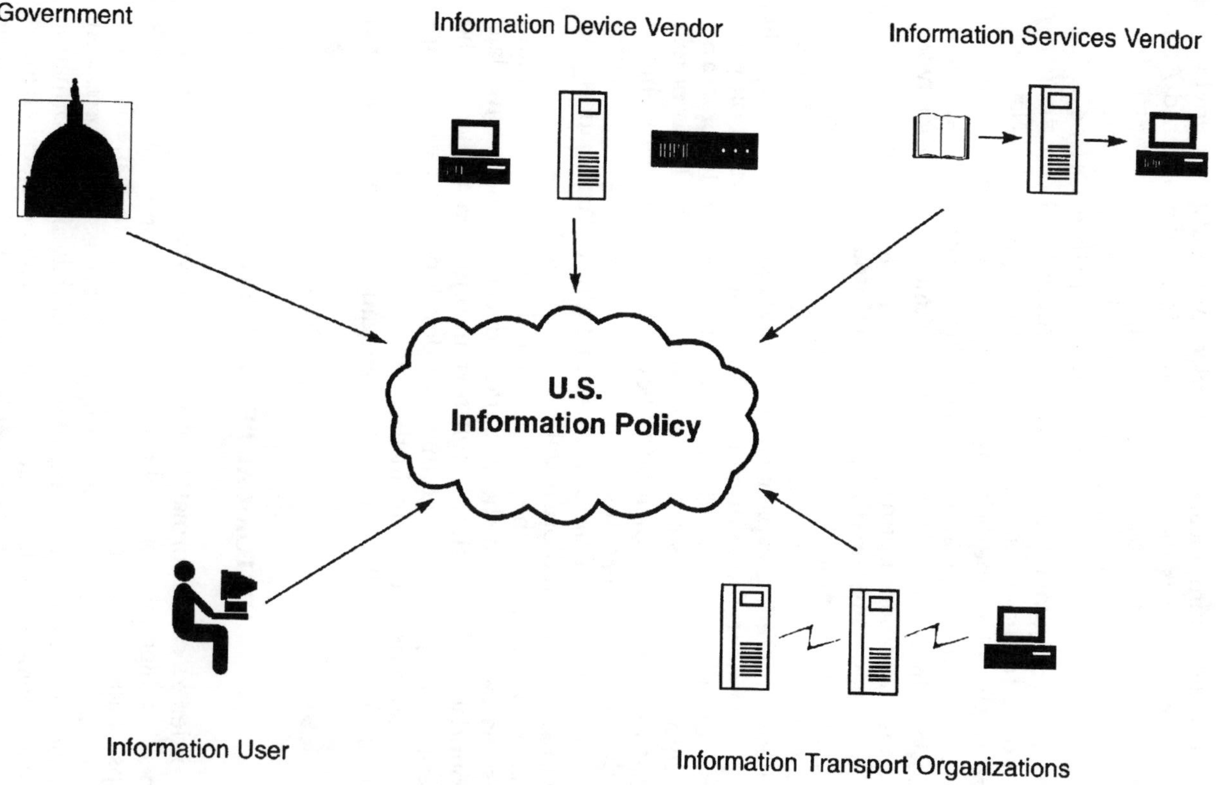

Figure 7.1 The U.S. Information Policy Generation

Initiatives Emphasizing the Medium

Full text information delivery has been driven by computing and telecommunications. The latter has to a marked extent controlled the medium. Defined broadly, it is the conduit that enables information to be remotely accessed. Interestingly, the policies have moved from centralized regulated communications providers to a deregulated environment. The 1934 Communications Act started the ball rolling by creating the Federal Communications Commission, which regulated telegraph, telephone and radio. This Act had major amendments added in 1952, 1960, 1978 and 1981 in order to accommodate new developments that dealt with television, satellite, and cable-based systems. American Telephone and Telegraph (AT&T) enjoyed the sole telecommunications monopoly. Starting in the 1960s, telecommunications, microelectronics, and computers became significant to the rest of the U.S. economy. Integration was needed, as well as developing innovative applications using the combined technologies. AT&T developed some applications to cater to changing industrial needs, but its main concern remained the telephone. In spite of its solid research and sophistication, AT&T could not meet the industrial demand for information technology based products.

Progressive divestiture of AT&T started in 1969 when Microwave Communications, Inc. (MCI) and other common carriers entered the long distance telephone market. A combination of factors that included Department of Justice displeasure with how competitors were treated by the local Bell Operating Companies and more demand for sophisticated services that AT&T could not meet, led to the disintegration of the Bell telephone system. Accordingly, Judge Harold Greene presiding, a consent decree was issued in 1982 for the company to divest itself of the local Bell Operating Companies (BOCs) so that all carriers had the same access treatment when their customers' calls originated or terminated at the local Bell Central Offices. Divestiture brought innovation and resourcefulness to the telecommunications environment. As mentioned already, MCI was one of the first U.S companies to offer e-mail to its customers. E-mail introduced a form of full text delivery that business and industry needed for fast and effective information delivery.

Deregulation took another turn in 1996 when Congress passed the Telecommunications act of 1996. (S. 652, Public Law 104–104) in February, 1996. This Act pulled all plugs and left all communications companies free to participate in whatever information activities they chose. The four major groups of players likely to take full advantage of the open competition are: long distance telephone service providers; local telephone service providers, cable television companies; and broadcast companies. There are bound to be mergers and acquisitions which may result in high-quality services and products. There will, however, be a learning curve when cable companies have to deal with data rather than images. Similarly, telephone companies, which have a lot of expertise in voice, will have to handle multimedia that may include full text online database development. Consolidation of the owner of the message and the medium all in one, may result in enhanced services.

This is an extremely comprehensive Act with far reaching consequences. A summary of the major provisions that affect full text information delivery is given as follows:

- Broadening the local exchange market. By declaring that no state or local government statute, regulation, or legal requirement can prohibit any entity from providing either interstate or intrastate telecommunications service, the 1996 Act makes all markets free for all.

- Retention of the universal service guarantee. The Act stipulates that the FCC and the states should ensure that universal service is available at just, reasonable, and affordable rates. Six universal service principles were stated and cover: quality service and affordable rates; access to advanced services; access in rural and high cost areas; equitable and non-discriminatory distribution; special and predictable support mechanisms; and access for public institutions.

- Removal of the MFJ restrictions for Bell companies. Modified Judgment (MFJ) of 1982 restrictions required that AT&T relinquish its local service operations. It restricted the regional Bell Operating Companies (BOCs) to local services and also banned them from information services, interregional services, and telecommunications equipment. The new law released BOCs from the most significant restrictions originally set by the MFJ.

- Entry of telephone companies into video services. The old communications Act barred telephone companies from providing cable TV service in their service territories except rural areas. The 1996 Act removed the ban and allowed the companies to get into cable TV and other video services.

- Entry of cable TV providers into telecommunications. Under the new law, cable TV companies may provide telecommunications services. This may be done with or without a cable TV franchise.

To the information user, the 1996 Act brought diversity of operators competing for the same consumer buck. While competition sometimes results in improved services is does not necessarily guarantee them. There will certainly be shake-outs among information providers and innovative services and products will emerge. The aggregate effect on the quality, quantity, and accessibility to available services remains to be seen.

Initiatives Affecting the Message

Highlights in the history of initiatives that affect the message include the Freedom of Information Act; The Privacy Act; and the Copyright Act. Technological development in cyberspace has made these legislative initiatives very complex indeed.

Freedom of Information Act

Enacted in 1966, The Freedom of Information Act (FOIA)—and amended in 1986—aims to set standards for the public to have access to government held records. FOIA is based on the principle that the public should have access to information held by the government except where confidentiality may be compromised when legitimate individual or governmental privacy concerns are involved. While FOIA protects the general public's right to know, with regard to government held information, there are several exceptions, with the overriding restrictions centering on national security and the individual's privacy when personal records are at stake. FOIA remains intact within the context of cyberspace.

Privacy Act

The Privacy Act of 1974 is comparable to FOIA in that it permits individuals access to Federal records that concern them. It may be used as a verification device to make sure that information held about oneself is accurate. Like the FOIA, the Privacy Act is applicable within cyberspace. With appropriate identification, individuals can now view and download full text information from some government agencies. However, with the increasing incidence of hackers and unauthorized access to organizational computers, there are restrictions imposed to protect government databases which may curtail the intent of this Act. In addition, protection of individual privacy provided under this Act faces tremendous problems when agencies have commonly shared databases. Strict safeguards must be undertaken, otherwise personal and private information can be accessed by criminal elements and used for blackmail and extortion.

Copyright Act

Finally, the Copyright Act passed in 1976 made comprehensive revisions to U.S. copyright law based on the 1909 Copyright Act. Basically, it strengthened copyright ownership. At the same time, it made it easy to copy under section 107 that defined what is "fair" copying. Section 105 of the Act bars the Federal government from copyrighting its publications and thus guaranteeing public access. The Copyright Reform Act of 1993 attempted to update copyright so that it is in line with technology. It simplified deposit for software by modifying the conditions of deposit. A short form was introduced that makes it easier to file. It also authorized the Registrar of Copyright to identify cases for which the purposes of registration and deposit may be served by deposit of descriptive material only. A written request by the owner of intellectual property may be accepted to defer deposit of copies later. The new provisions may safeguard ownership and at the same time encourage deposit, since the whole software code for a package or a CD-ROM database product may not be deposited at the Library of Congress.

In spite of the 1993 legal provisions easing depository regulations, the Act emphasizes the importance of the deposit obligation for all copyright holders to deposit material at the Library of Congress. The new legislation gave the Librarian of Congress rather than the Registrar of Copyrights the authority to enforce the depository requirement. As part of this responsibility, the Library of Congress must establish an electronic database of all deposits. The database must be made accessible through one or more international information networks. While the copyright law is still intact and applicable to cyberspace, enforcement continues to be problematic as software and other copyrighted full text material can be easily pirated in cyberspace.

Initiatives Impacting Access to Information

The Depository Library Act of 1962 consolidated existing statutes that provided for distribution of government documents. It increased the number of legal depository libraries throughout the United States. They were designated as "legal deposit" because they were entitled to receive free government documents. The main goal of the depository system is to make U.S. government documents easily accessible to the general public and to ensure that availability is continued in the future. In effect, the system guarantees free access to whatever emanates from the government that is not classified as inaccessible due to national security or individual privacy considerations.

The Superintendent of Documents at the Government Printing Office (GPO) is responsible for distribution of documents to the designated libraries. In return, libraries must open their doors to the general public so they have access to these documents, if nothing else. There are well over 1,000 libraries that are designated legal depositories, scattered throughout the U.S.

What information is distributed? All documents published by the government. This is true whether they originate at the GPO or from other printing sources like federal agency printing departments. Exceptions include: documents generated for administrative use only, classified material due to national security requirements, or documents with no educational value and are thus unlikely to be of use to the general public.

A recent development that updates access to government information was the GPO Electronic Information Access Enhancement Act of 1993. The Act was enacted in the spirit of free public access to government information. It requires the GPO to provide the following services in order to promote public access:

1. Maintain an electronic directory of federal electronic information.
2. Develop a system of online access to the *Congressional Record*, the *Federal Register* and other publications distributed by the GPO.
3. Develop an electronic storage database (facility) for federal electronic information.

This Act responds to a point raised at the beginning of this chapter which states the depository system has to be revamped in view of the information superhighway. According to the terms of the Act, legal depository libraries will have free access to electronic databases. It establishes a five-year multi-million dollar project aimed at bringing the GPO to cyberspace. Other than going through the depository system, other modes of access will entail a fee-based, incremental dissemination incurred by the GPO.

Prior to the Electronic Information Access Enhancement Act of 1993, a Paperwork Reduction Act (PRA) was passed in 1980. The basic objective of PRA was to reduce the paperwork burden caused by federal regulation requirements. It directed the Office of Management and Budget (OMB) to oversee the development of federal information policies governing practice in all Federal agencies. Policies would also cover information collection and reporting that is required of the private sector in the hopes of reducing the paper burden on the private sector. In order to fulfill its obligation, the OMB issued Circular A-130 in 1985 entitled, "The Management of Federal Information Resources." It provided the general policy that agencies should follow together with the rationale. Six principles underlying federal information management follow:

1. Information is an economic resource. Since it has cost and value, it must be managed like all other resources.
2. Information has a life cycle. For example, presidential orders issued 20 years ago may have lost their potency and relevance to society. Government information is segmented according to source whether legislative, judiciary, or executive. At the same time, it is related and links are needed.
3. For the sake of efficiency, especially cost reduction, agencies must use new technology to process information.
4. Accountability and management of information resources must be decentralized due to the size and diversity of federal agencies.
5. The role of the central administration (OMB) is to initiate, develop, and implement policy directives to guide agencies in their information generation and dissemination.
6. With regard to individual privacy and national security, federal information requires different considerations in contrast to other types of information.

Several critics have pointed out that the PRA had a dampening effect on access to federal information, since it generally reduced the quantity of hard copy information disseminated due to economic reasons. In addition, it encouraged more classification for security purposes of government information sources in the gray area. On the other hand, principle three laid the foundation for government participating in cyberspace, as it encouraged development of electronic full text databases within the agencies as well as access systems. Within the last decade or so, federal agencies have developed several systems for electronic collection and dissemination of information to the general public.

A hybrid of initiatives that include the medium and the message embody legislation concerning the development of high performance networks and applications. They have resulted in direct Internet funding for research via the National Science Foundation, maintenance of the backbone National Science Foundation Network (NSFNET), research for establishing digital libraries, and healthcare through the National Institutes of Health and the National Library of Medicine. The cluster of legislative initiatives include: The High Performance Computing and Communication Act of 1991, the original Gore Bill; The National Information Infrastructure Act of 1993; and The National Competitive Act of 1993. This category of legislative effort has brought the U.S. government to the forefront of providing leadership in laying the foundation for electronic full text delivery by directly supporting the development of the national backbone and supporting networking research and demonstration projects.

THE NATIONAL INFORMATION INFRASTRUCTURE

The Concept

The Internet proper was a spin-off of research and development stimulated during the Reagan-Bush Administrations of the 1980s. While the Republican administrations of the 1980s were keen on supporting research, they did not encourage government initiatives exploiting the enabling technologies. Some critics have argued that this apparent lack of interest was due to a strong communications lobby that worked hard to fend off government interference. Defense research was heavily funded in the 1970s and 1980s, but statutory provisions barred this research to be diverted to civilian applications. It is ironic that the Internet as we know it today owes its beginnings to defense research effort.

Vice President Al Gore (then a Senator from Tennessee) initiated the larger picture by coining the concept of a National Research and Education Network (NREN). NREN became the intellectual embodiment and catch-word of government policy development and involvement in molding cyberspace and, ultimately, the direct promotion of electronic full text delivery. He strongly projected the NREN concept into U.S. politics as a stepping stone for launching a series of viable policy initiatives. With the passage of The High Performance Computing and Communication Act in 1991, the concept was formally adopted by the U.S. polity.

As a concept, NREN had a number of functions. First, it served as the rallying point to mobilize mass national support for a national information network. A national research and educational network appeared to have something for everybody. It sang the tune of spreading educational programs that would affect almost everyone. Second, the research effort would also benefit the whole nation even if this might mean a trickle-down policy. Finally, the business and industrial entrepreneurs saw the grand project as an opportunity-creating environment. With the whole world embracing cyberspace, national and international business enterprises would be greatly enhanced. Electronic full text delivery would thrive. NREN aimed to:

1. Promote U.S. competitiveness on the world markets (by speeding up business transactions).
2. Provide faster and expanded access to super computing.
3. Minimize spatial restrictions for researchers (less travel needed, reduction in cost of correspondence, and reduction in project completion time).
4. Promote collaborative research (thorough the use of e-mail and file transfer systems for uploading and downloading).
5. Improve the quality and quantity of education by simplifying access to resources.
6. Promote and enhance technology transfer by improving speed and mode of transfer.

These perceptions have been sustained throughout the information policy initiatives' planning and implementation.

The Commitment

Establishing official organs for policy generation marks the commitment to promoting and eventually implementing a policy initiative. A High Performance Computing and Communications Program (HPCC) was established as a result of The High Performance Computing and Communications Act of 1991. It was incorporated in the National Information Infrastructure (NII) proposals. HPCC is an interagency program, with about $1 billion in federal support, that coordinates federal funding for research and development at universities and colleges. The research is directed towards developing more powerful computers, faster networks, and better software.

On September 15, 1993, the federal government issued the ambitious *National Information Infrastructure: Agenda for Action*, in which nine guiding principles and objectives were identified as follows:

1. Promote private sector investment.
2. Extend the "universal service" concept to ensure that information resources are available to all at affordable prices.
3. Act as a catalyst to promote technological innovation and new applications.
4. Promote seamless, interactive, user-driven operation of the NII.
5. Ensure information security and network reliability.
6. Improve management of radio frequency spectrums.
7. Protect intellectual property rights.
8. Coordinate information network development efforts with other levels of government and with other nations.
9. Provide access to government information and improve government procurement.

The NII document was issued by the NII Task Force, chaired by The Secretary of Commerce, Ron Brown. As part of the development chronology of government information policy initiatives, it is important to note that the task force oversaw three working groups: government information, intellectual property, and privacy. It is interesting to note that the Intellectual Property Working Group was chaired by the Commissioner of Patents and Trade Marks of the Department of Commerce. It sought industry opinion on how to protect copyrighted material.

Implementation

The Government Information Locator Service (GILS) that was proposed by the OMB and the Information Policy Working Group of the NII Task force drew attention to the need for developing a system for "actual" identification, access, and delivery of full text government information. The focus on federal agencies was particularly pertinent. The National Technical Information Service (NTIS) already has a system similar to the GILS proposal called

FEDWORLD. More than one hundred online, federally operated systems are covered under this NTIS system. On the other hand, the GPO has the obligation to provide an equivalent service under the GPO Electronic Information Enhancement Act of 1993. At the time of this writing, the apparent conflict had not been resolved, but the GILS service as proposed is invaluable for accessing full text government information.

NII is a complex government information policy initiative that is being implemented piecemeal. Several actions are taking place on many fronts. For instance, several programs have been funded within the spirit of the principles outlined in the *National Information Infrastructure: Agenda For Action* document. Many of the projects are aimed at demonstrating the feasibility of networked services or products using the Internet or regional networks. One of the major projects was the NSF-funded project to make EDGAR, the Securities and Exchange Commission (SEC) databases available over the Internet. New York University's Graduate School of Business got the grant. This is a prime example of full text implementation, as it will include corporate information filings as provided to the SEC, including graphics and other illustrations. The general public, researchers, and business corporations would access this information source at a relatively lower cost than sources provided by commercial organizations.

INFORMATION HIGHWAYS FOR THE 21ST CENTURY

Effect of Policy Initiatives on Network Development

Since the High Performance Computing and Communications Act of 1991, federal seed money for networking has been forthcoming through the National Science Foundation and other agencies. Much of the money has been in the form of grants and contracts. Regional network development has been funded as well as the access to networks developed by organizations.

Public access to networked information has not been solved and is perhaps one of the most difficult issues. Access to networks through legal deposit libraries might be an option, but they may be far from the average "public" information user. In effect, despite the "grand" picture of the NII, the average American will have trouble benefiting from NII. The "public lobbies" must continue their efforts to draw attention to this imminent anomaly. Otherwise, full text information delivery systems will be a privilege of the elite who can afford it as provided by a few very large multinational corporations.

International Networking Implications

The international U.S. network agenda is a classical multi-lateral or bi-lateral relationship. The three organizations that have dominated the rules of the game at the bilateral level are: International Telecommunications Union (ITU), International Telecommunications Satellite Organization (INTELSAT), and International Maritime Satellite Organization (INMARSAT). The process of achieving consensus at these organizations is becoming more difficult as the stakes associated with information become more significant to national economies. The State Department has carried the banner with technical input provided by the Federal Communications Commission (FCC) and the National Telecommunications and Information Administration (NTIA).

Bi-lateral and regional developments have continued to reflect the U.S. and the "other" nation's interests. Some of the countries that have benefited from these arrangement have included the United Kingdom, Canada, Japan, and Mexico. Areas covered under the bilateral agreements are information policy issues not covered by the multilateral arrangements. For full text information exchange to be carried on between two countries, international telecommunications carriers must agree to cooperate to synchronize operations. Much of the underlying full text information exchange issues are transparent to the information user. It suffices to say that multilateral or bilateral agreements allow the full text information exchange that we take for granted to take place. Through these agreements, the information super highways for the twenty-first century are gradually being molded.

References

Bender, David R. "A Strategy for International Information Policy." *Libri* 43 no.3 (1993): 210–231.

Brown, J.S. "Research That Reinvents the Corporation." *Harvard Business Review* 108 (January/February, 1991): 330–339.

Burger, Robert. *Information Policy: A Framework for Evaluation and Policy Research.* Norwood, NJ.: Ablex Publishing Corp, 1993.

Chartrand, R. L. and J. Bortnick. *Information Policy and Technology Issues: Public Laws of the 95th Through 101st Congresses, CRS Report 91–569 SPR, Congressional Research Service.* Washington, D.C.: Library of Congress, 1991.

Committee on Applications and Technology. *What It Takes to Make It Happen: Key Issues For Application of The National Information Infrastructure.* Washington, D.C.: Information Infrastructure Task Force, 1994.

Galvin, Thomas J. "Leadership in Legislation and Public Policy Development: The Case of the American Library Association." *Library Trends* 40 no.3 (1992): 431–56.

Garrett, J.R. "Digital Libraries: The Grand Challenges." *Educom Review* 28 no.4 (July/August 1993): 17–21.

Hernon, Peter. "Government Information Principles." *Government Information Quarterly* 8 no.4 (1991): 393–399.

Hernon, Peter and Charles McLure. *Public Access to Government Information: Issues Trends and Strategies.* Norwood, NJ: Ablex Publishing Corp, 1984.

McMundo, George and Evelyn Simpson. "The White House Files." *Journal of Information Science.* 20 no.5 (1994): 305–313.

Newberg, Paula R. *New Directions in Telecommunications Policy.* Durham, NC.: Duke University Press, 1989.

Sprehe, J.T. *The Year in Federal Information Policy: Events and Commentary: 14–15, (IDC Government Report),* Washington, D.C: Government Printing Office, 1993.

U.S. Congress, Office of Technology Assessment. *Making Government Work: Electronic Delivery of Federal Services.* Washington, D.C.: Government Printing Office, 1993.

The White House. *The National Information Infrastructure: Agenda for Action: 3, by the Information Infrastructure Task Force.* Washington, D.C.: The White House, 1993.

Wicks, Wendy ed. *Government Information and Policy: Changing Roles in a New Administration.* Philadelphia: The National Federation of Abstracting and Information Services, 1994.

8. Electronic Full Text Information Delivery Services in Cyberspace

The need for delivery of full text information to a user's desk is perennial and common to all sectors of society, whether corporate or public. As explained in earlier chapters, full text means information that has text, images (graphics), and in some cases, sound. Multinational corporations like IBM, ICL, Digital Equipment, GTE, and Siemens have proprietary networks which transport full text on an intercontinental basis. With these networks, proprietary graphic designs for new products which are developed at a location on one continent can be electronically transmitted to corporate manufacturing plants on another continent. Information intensive, but not necessarily high-tech companies, also lease-high speed telecommunications lines to accommodate their electronic full text delivery needs. Organizations in this category include banking institutions, insurance companies, hospitals and health management agencies, to mention a few. Similarly, many U.S. federal and state agencies have access to high-speed telecommunications lines that facilitate full text information delivery. Asynchronous Transfer Mode (ATM) has been touted for some time as one of the ideal protocols to use for general purpose electronic full text information delivery. As discussed in Chapter Four, the Internet also has a number of protocols that accommodate full text as defined in this chapter.

As information technology becomes relatively more affordable, electronic full-text information delivery technology is experiencing a trickle-down effect. Starting from the giant multinational corporations, which had monopolized it for more than a decade, smaller organizations are beginning to use it daily in their operations. Information centers and libraries are fairly new in the electronic full text delivery game. They have special problems since much of the information they have traditionally handled is another organizations information. While they predominantly dealt with hard copy, they were shielded by the U.S. Copyright Act under the "fair use" clause. Even when technology becomes fully affordable, user-friendly, and accessible, information centers and libraries will still be haunted by the issues related to intellectual property.

Proprietary networks aside, much of the information needed by the average user emanates from some publishing agency or enterprise that has rights to its intellectual content. Inevitably, such restrictions raise issues of legitimate reproduction—manually or electronically. These somewhat special circumstances are particularly important to information centers and libraries that serve the general information user. The information generation and disseminating process may be envisioned as a multi-phase chain (see figure 8.1) with the following broad categories:

1. Owners of intellectual property who have published copyrighted creative works (e.g., authors or publishers).
2. Online data services—mainly commercial service vendors (e.g., LEXIS/NEXIS, Knight-Ridder, Dow Jones, etc.).
3. Information and information access providers—information brokers, information centers, and libraries.
4. Clients—information end-users.

Information centers and libraries have traditionally been a third party in the information processing and dissemination chain. They are caught between the owners of intellectual property and online databank developers on one hand, and the end-user on the other.

This chapter focuses on the issues associated with electronic information full text delivery that involve third parties which provide multi-purpose information access for public consumption. For the purpose of our analysis, a distinction will be made between electronic full text information delivery and electronic full text document delivery. The former will be used as an all-embracing concept that includes all forms of delivery, whether the information in question ever existed in hard copy or not. On the other hand, electronic full text document delivery is used with the connotation that emphasizes the fact that there has been a hard copy form of the original "document" which was digitized and made available in machine-readable form. In many cases, the digitization process is provided by an information capture device that may digitize text, sound, graphics, or images.

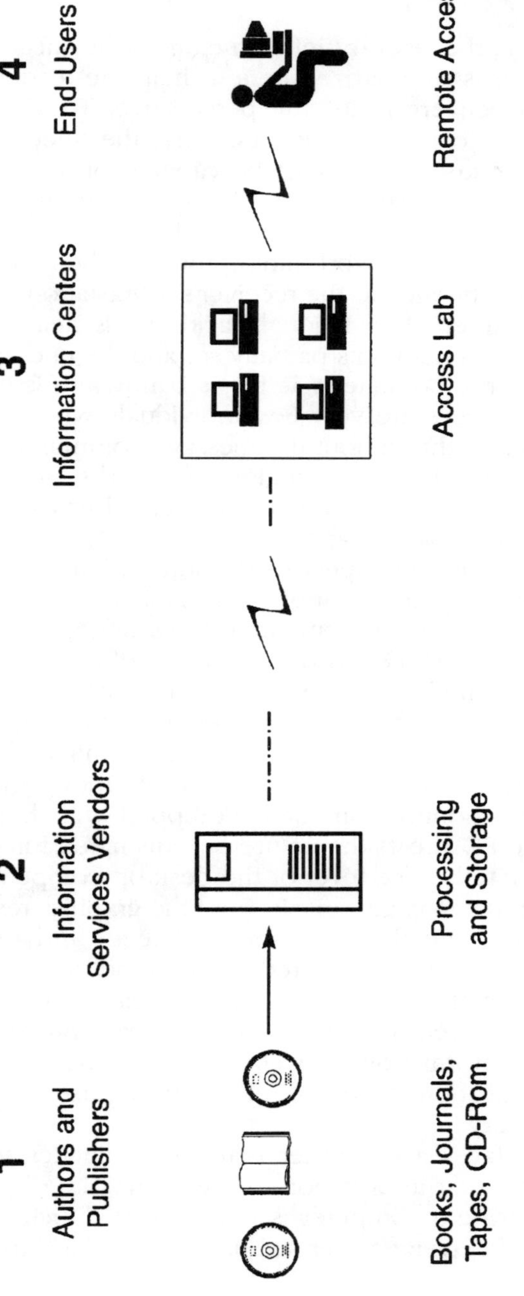

Figure 8.1 Information Generation Chain

THE RELEVANT INFRASTRUCTURE

In essence, the relevant infrastructure is the substance of much of this text. It is summarized here to help the reader see full text information delivery in its true perspective. To achieve effective electronic full text information delivery, the telecommunications part of the infrastructure must be capable of accommodating it. This inevitably translates into a discussion bandwidth size as explained in Chapter Two. Ideally, high bandwidth should be available from the sender's workstation, through the national network, via regional networks, to the receiver's workstation. As mentioned in earlier chapters, the national backbone is nearly all fiber optics, the regional networks partially so, and the end-users so-called "last mile" is the bottleneck as far as bandwidth is concerned. In other words, there are very few individuals who are guaranteed high bandwidth throughout the message communication process. Many go at a snail's pace to regional networks, improve their rate of transmission at the regional, and get a burst of speed at the national information network.

Another significant aspect of the infrastructure that affects the full text electronic information delivery is the basic sending and receiving device. In telecommunications lingo, it is the data terminal equipment (DTE) while for the rest of the information community it is simply the terminal. In the last few decades, the "terminal" has come to mean an intelligent PC. Its intelligence lies in the fact that while the early types of terminals had very little storage buffers and virtually no local applications, the modern PC is a fully functioning computer. Equipped with high-level functionality, the PC affects the nature of transmitted full text. In other words, the intelligence level of the desktop computer or the network server, the storage capacity, and the graphics resolution quality will determine the sophistication level and quality of the electronic information delivered. At the time of writing, there are several brands of PCs on the market that can be used for the type of delivery covered in this section. The main problem is that the bulk of the information user clientele does not have the equipment sophistication to benefit fully from the use of full text information delivery.

User interfaces are another issue that militates for or against accepting full text information delivery. In spite of graphical user interfaces, using a computer is much more difficult than using a typewriter. The average user still has to work hard to reach a level of comfortable use on most systems. Even the ones which are fully icon-oriented need a time investment to achieve ease of use. Systems developers still have a long way to go in achieving truly user-friendly systems.

Finally, user computer literacy as a deterrent to effective use is a temporary condition, as U.S. children are now introduced to computers in kindergarten. In a nationwide study that was done by the author and reported in Chapter Five, information professionals seem to think that computer literacy will be a minor factor in adopting full text delivery systems by the end-user. Clients that come from school systems and higher education that have emphasized the computer as a survival tool will have a life-long appreciation of full text delivery. For the time being, however, lack of adequate computer literacy is still a drawback for the average information user.

CHOOSING ELECTRONIC DELIVERY

The late 1990s have been a period of relatively slim budgets for information centers and libraries. It is unlikely that funding for information services will be restored to the levels enjoyed in the previous two decades. In response to information on demand under tight budgets, several organizations are resorting to alternative means of providing full text service to clients. Electronic full-text information delivery is becoming a viable alternative for a number of reasons.

Led by Dialog Information Services, now a Knight-Ridder subsidiary, database service vendors have for three decades provided online citation identification services. During this period, the information user has been served with miles of printouts of citation-based sources. He or she gleefully accepts the results of a computer search with the hope that much of the generated universe of relevant information will be available from the in-house information resources. The greatest disappointment has always been the lack of these resources, either in-house or the immediate regional resources. End-users have suffered this inconvenience since there were not many other alternatives until the early 1990s. As all information professionals know, the ultimate in information seeking is the receipt of the substantive article or report rather than an index to or abstract for it. Electronic document delivery services have come to the rescue, although many of them are still crude. The rationale for resorting to electronic information delivery is discussed in the following section.

Rationale

Choosing electronic information delivery is not simply a better option, there are some fundamental reasons why it is bad business not to use it. As most analysts of scholarly publishing acknowledge, there is a crisis in journal publishing. Several information centers and libraries spend more than fifty percent of their materials budget on journals. Yet, within the last ten to 15 years, journal, indexing, and abstracting prices have almost doubled.

Also, the rate of journal publishing has been very high in the last two decades. For instance, the price of chemistry and physics literature has doubled in this period. Thus, there are more potentially relevant science and technology titles on the market than 15 years ago. This phenomenon makes it very difficult for the scientist to keep abreast of his or her subject. At the same time, it also makes it hard for the information provider to give effective service to the research community using hard copy. Alternative choices to hard copy became imperative.

Information centers and libraries in all sectors of society are undergoing re-engineering and downsizing. The emphasis is on doing more with less, particularly within the personnel and material resources budgetary items. The resultant cancellation of both books and journals has called for using the principle of access, rather than ownership, of materials. Put in another way, information centers and libraries have to move away from a policy of materials acquisition "just in case" some one comes for it to a policy of getting material "just in time" for someone who has made a "live" request. This negates building large collections of material for an invisible and unpredictable future researcher. Full text information delivery systems are ideal for providing material in this new environment.

There has been an upsurge in client demand for full text information in lieu of citations. In many cases, full text is demanded because items are not available in the local information center or library. User appetites have been increased by the use of CD-ROM databases which are searched locally with no time limits. In addition, databases loaded on online catalogs, have generated a high demand for material outside the institution. In corporate environments "full text information on demand" rather than citation based information is more of the norm than the exception due to the need for much of the corporate information to be part of the decision support system. With international competition in the backyard in addition to overseas markets, U.S. corporations' need for instant information has become greatly exacerbated.

Very few, if any, information users ever read a journal from cover to cover. It is normally a selective reading, picking articles relevant to a current interest, be it research-oriented or otherwise. It may be argued that another information user may need the remainder of the journal. Depending on user interest, which should be regularly monitored, selective acquisition of partial journal information may be justifiable. Subscriptions to heavily used journals may be maintained while those for less-used ones can be replaced by a selective acquisition policy on demand.

Finally, technology is a major factor in prompting new enhanced electronic full text services in many ways. In the first place, national, regional, and local networks make retrieval feasible. Many information centers and libraries are units in institutions that use full text delivering networks for other functions. Adding, sending, or receiving text that emanates from the information center, rather than from computer aided design (CAD) or computer aided manufacturing (CAM) units, may not be that much of a burden to the institutional network. Also, since many of the users already upload and download full text material from networks, their expectations would change *vis-à-vis* the inhouse information center or library. In addition, many of them may be the targets for advertisements for full text delivery, either directly or via the information center. All these considerations would change their attitude to what they should be entitled to from their local information center or library.

Applicable Technologies

Beyond the general considerations for providing a conduit or even a sending or receiving terminal, there are specific technologies for full text delivery. Within the last two decades, experiments with scanning and digitizing text, images, and sound have resulted in the development of commercial systems now available on the market. Given the necessary equipment, it is now possible to generate and transmit multimedia documents over the information networks. It is assumed that an appropriate bandwidth exists.

The capacity to display and reproduce a transmitted multimedia piece depends on the ability of the video screen to reproduce the graphic, and the sound mechanism to accompany the display. This translates into the digital storage, sound card, and speakers combination. Much of this technology is being used already for games and other applications. Clarity of reproduction will depend on the quality of the product used on either end of the network.

Storage at both ends concerns *temporary* buffer storage as the piece of creative work is being transmitted or *permanent* storage when the material is inhouse. Two techniques are available on the market to facilitate processing high-volume large files. One is the use of compression/decompression techniques. Another is the use of high powered (bufferred) screens and general RAM space. Video RAMs are now at megabit levels for PCs. The resolutions are improving to make the redrawn pictures much better at the receiving PC with higher refresh rates. Similarly, standard RAM storage is expanding and hard disk storage at the gigabit level are already on the market. In the final analysis, the final product depends on how the subsystems have been assembled to respond to local information needs.

Finally, if the downloaded full text material is to be sent to a printer, other considerations come into play. Color and graphics capability will again depend on the sophistication of the printer available. There are presently several brands of printers that can print much of the text and graphic type of full text as envisaged by this text.

Types of Generic Systems

There is an explosion of full text information delivery systems on the various networks. Some of them are conventional and have been on the market for some time. Others are new and may not even be thought of as sources of information of first resort. As this is a rapidly growing area, the list is meant to be indicative rather than comprehensive. Six generic systems have been identified for a brief discussion.

The CD-ROM (compact disk read only memory) has been on the market for over ten years. Comparable to the audio CD, it has very similar characteristics of durability, very high density storage, and interactive direct access. Many of the CD-ROM products with full text information are by products and subsets of online systems. While they provide unlimited searching for information that they may contain, they are not as up-to-date as their online counterparts. Until recently, many of them were used only in a stand-alone mode. "Juke boxes" and networked CDs are now available and used by information centers and libraries.

There are also locally loaded *electronic full text sources.* In this context, local loading means that an organization has created its own inhouse databank of several databases. Locally loaded information sources fall into two broad categories. In the first place, there are resources which may be on campus or institutional issues. These may be available to both institutional and external users. In the second place, there are sources that are copyrighted and obtained from other institutions via subscriptions of one type or another. Among the common examples are: ABI/INFORM, MEDLINE, and Harvard Business Review. The second category normally requires some form of identification before searching. The main advantage of these sources over the CD-ROM is the fact that they can be searched using one interface, which is usually the local online catalog that may be part of the campus-wide information system (CWIS). A disadvantage is that they are not as up-to-date as the direct online versions from the vendors. Currency will ordinarily depend on the vendor's update policy.

Electronic journals (EJs) have been introduced to the information arena within the last few years. A small number of them are refereed, which should add some respectability to the material they publish. They are offered through the public networks and a few of the commercial online services. While electronic journals may be ideal as a mode of delivering full text information to bona fide researchers, the top exclusive journals are yet to appear in electronic form. Subscriptions to the regular journals have not as yet been affected by the advent of EJs.

Perhaps the most sought after system by information centers and libraries is *commercial online interactive digital storage and access.* There are two components to this mode as it is gradually developing. First, there is the table of contents (TOC) database which, depending on the system, is derived from a finite set of journals that a given vendor has chosen to put in the system they have on the market. TOC databases are no more than lists of citations or finding tools, which may or may not be associated with full text delivery service. Examples include Faxon Research Services' Faxon Finder and CARL's UnCover. TOC allows users to identify the exact journal article needed so that it may be ordered from a document supply agency. The second component is the document delivery service, which provides the actual articles or other documents identified. Examples in this component are CARL's UnCover2 and Faxon Research Services' Faxon Xpress. The two components may or may not be part of the same system. After identifying the article(s) needed, the user may connect to another system that supplies the actual document. One of the main problems with existing systems is the very limited coverage with regard to the number of journals represented by any of the systems on the market.

Local freenets form another generic system that may be accessed for electronic full text information. They are collections of electronic databases on local city or county information. Among the various aspects usually included are: local restaurants, places of interest to visitors, and directories of local officials. Cleveland, Ohio established one of the first freenets on the Internet. They are 'free' because one does not need an ID number to access the information they contain. Information center personnel should keep an online list of all freenets which may have special features of interest to the organization they serve. Since sources of this nature are easily accessible to the end-user, the user should be encouraged to have direct access. Freenets have factual, local information which may include statistical tables that researches may need on the particular site. They are invaluable when one wants to travel to an unfamiliar location. Instead of going through published maps or guides, information about a given place can be easily obtained from a local freenet. If need be, such information may be downloaded and taken as travel documents for future reference. The most obvious advantage of using freenets over published guides is that they tend to have more current information than the guides.

Finally, electronic bulletin boards (EBBS) are another possible generic source of full text information. Identifying the relevant board may not be as easy as one might expect. Some bulletin boards have been expanded in coverage since the initial introduction. Perhaps the most difficult aspect of EBBS is the fact that they span private and public networks. The most relevant may be on a private network that requires subscriptions both to the general network and the bulletin board itself. They are, however, another nonconventional category of sources of full text information that might be of use to any organization.

Types of Vendor Applications

In an attempt to circumvent the "miles of citations" dilemma, the information arena is gradually being flooded by a large number of practical commercial full text document delivery systems—see Appendix A. They are in effect two stages of operation. The first stage is item identification provided by a table of contents database. At the second stage is article identification and supply. Due to the need for expediency in a semi-developed environment the two stages are not necessarily performed by the same organization. Stage two is implemented in four ways.

First is the publishers initiated service. An institution may subscribe to a CD-ROM database of digitized images of full articles from a selected set of journals. ADONIS is such a service founded in 1991 by a consortium of publishers. With its main office in Amsterdam (Holland), it was founded by Blackwell Scientific Publications, Elsevier, Pergamon, and Springer Verlag. Over forty publishers now participate and contribute full text articles from over four hundred journals. The database is available locally in the library/information center with updates of new articles provided by ADONIS at regular intervals. A main advantage of this arrangement is the fact that the materials are locally available and can be accessed, retrieved and printed off whenever the information center is open. A major disadvantage is the fact that the list of journals is very limited.

Second is a consortia or network initiated service, with cooperating libraries that hold the actual articles. CARL's UnCover is an example that includes citation identification. CARL's UnCover2 links the user to a cooperating library that contains the journal article. CARL is a consortium of libraries based in Denver, Colorado. When there is a match between the citation and the article, a cooperating library scans the article and transmits it to Denver where it is relayed to the requester. Both UNCover and UnCover2 are online services available via the Internet or direct dial access.

Third, is the online vendor's full text delivery. Knight-Ridder (formerly Dialog Information Services) is an example that provides direct downloading from a selection of full-text databases contained in its databank. LEXIS/NEXIS also provides full text legal resources under LEXIS and general material under its NEXIS component. A main advantage of these systems is that delivery is immediate when one connects to a full text database, for instance Harvard Business Review on Knight-Ridder. The disadvantage is that the journal selection is dictated by the vendor and may not match the user's interests and information needs.

Finally, the general commercial and other category includes an amorphous group. Article Express International Inc. is a unit of Engineering Index, and provides articles identified via Knight-Ridder. Delivery to the user can be electronic via Internet or by manual systems like Courier. There are a number of varieties in this category: some have their own databases of articles while others use other organizations' sources.

Criteria for Evaluating Electronic Document Delivery Systems

In order to select a service provider, a number of criteria must be used. Among the most common ones are first, the subject content. Before any service is selected, its scope with regard to subject content must be ascertained. An organization's information need determines whether a given service provider will be appropriate.

Second, compatibility with existing systems that an organization uses helps system maintenance staff. Networked CD-ROM are now common and many information centers and libraries are using them. If CD-ROM databases are to be loaded on an existing LAN, care must be takes to ensure that software that came with the CD-ROMs is compatible with the network. Locally loaded databases, which are loaded on the system that runs the OPAC are just like any other data and compatibility is relatively easy to achieve. The NOTIS system is a good example that incorporates such databases and allows users to search them using its general search engine.

Third, ease of installation and maintenance is important. This may depend on whether the organization has substantial number of inhouse technical staff. Some systems need very little maintenance and with a few minutes of vendor demonstration local staff will take over the maintenance with little difficulty. Others are complex and may need technical help given by the vendor at installation.

Fourth, it is important to know whether the system will use dedicated equipment. Current trends show that in modern information systems applications design, common equipment is used. Both hardware and operating systems are usually selected from popular brands. Operating systems like DOS, Windows or Windows

95 are popular among developers. With regard to hardware, IBM and its compatibles as well as the Apple computers are some of the popular brands. But some system designers do still chose less popular hardware brands.

Fifth, payment options is another criteria that many information professionals look for in a system. Does the vendor accept credit cards? Does it offer bulk discounts? Some libraries and information centers have banded together in regional networks and purchase service from OCLC. Similar arrangements may be acceptable by other vendors on the market.

User access to the system forms another criterion for selection. Some systems appear to be primarily designed for information professionals rather than end-users. Unless the end-user is prepared to invest a lot of time in learning the system he or she may find it difficult to use. In addition, many users might prefer to search at home or office and such a service must be provided.

How large is the vendors installed base? Usually the larger the user base the level of user satisfaction of the user group. It is customary for vendors to suggest current users who may be consulted for reference. In addition, most current users are willing to volunteer comments on the vendor on request by prospective customers. Access to the Internet has enabled individuals to belong to electronic bulletin boards. Vendor evaluations are often done for individuals on request or simply volunteered by members.

Finally, how flexible is the system? This may be with regard to the internal error correction, for genuine accidental mistakes. If a user makes an inadvertent error, does the system provide correction without starting all over again? It may also be with regard to increased volume of users to be served. Flexible systems will stand minor modifications to accommodate changes in usage patterns. Most site licensing fees depend on usage patterns and the information specialist subscribing to a document delivery service should consider how such fees may be affected by varying usage at his or her site.

Cost Related Issues

Relative novelty of full text information delivery systems makes their cost fluid. There are, however, two general and rather broad categories. One group whose membership varies requires a lump sum which may be called a subscription or a deposit. Prices for deposits/subscriptions range from hundreds to thousands of dollars. For instance the Information Access Company asks $2,000–45,000 while UnCover will accept a $100 deposit.

Checking individual pricing schemes does not reveal any consistent pattern. Generally, the total price comprises a per page fee, a delivery charge and any copyright clearance fee. Cost statements may not specifically mention royalties to publishers, but most services assume the responsibility and build it into the overall cost.

Constraints

Realization of full text electronic information delivery beyond the fax is being achieved by several organizations. Third party institutions (including information centers and libraries) operate within constraints, many of them external. The concept of third party is used since they operate between the information providers and the end-user. As technology matures and becomes user-friendly, many of the obstacles will be minimized.

First and foremost, technology has not spread far enough to allow the average end-user to access and manipulate the systems although many of them are geared to end-user application. On the general level, the telecommunications conduit is almost ready, but the access devices are still inaccessible to the average user. The main drawbacks are price of equipment and varying degrees of computerphobia.

The second and most intractable obstacle is the copyright issue. While many electronic document delivery systems and services do take care of the copyright problem, some of the law's provisions grow murky when applied to the Internet or other networks and non-commercial organizations.

Third, it may be well and good to espouse the mechanics of the full text information delivery systems. How about the quality of the material delivered? Some analysts that have examined quality have questioned the infomation's usefulness when quality is questionable. The garbage in garbage out or GIGO maxim is applicable to some databases. Unfortunately, in many cases, the end-user has no yardstick to evaluate product quality.

Finally, how liable are the document delivery systems for the material they peddle? Given that information is normally used for critical decision making, should providers of defective information be liable if bad decisions are made because of it? A few liability cases have been decided with regard to information brokers and large information processing firms. Growth and maturity of the information business will certainly bring more law suits for malpractice or negligence.

References

Basch, Reva. *Electronic Information Delivery: Ensuring Quality and Value.* Aldershot, UK: Gower, 1995.

Bennett, Valerie. "Electronic Document Delivery Using the Internet." *Bulletin of Medical Library Association* 82 no.2 (April 1994): 163-167.

Bluh, Pamela. "Striking a Balance: Document Delivery in the Nineties." *Law Library Journal* 85 (1993): 599-608.

Campbell, Robert. "Document Delivery and the Journal Publisher." *Scholarly Publishing* (July 1992): 213-221.

Gassaway, Laura N. "Document Delivery." *Computers in Libraries* 14 no.5 (May 1994): 25-32.

Jackson, Mary E. "Integrating Document Delivery Services With Electronic Document Delivery Technologies." *Law Library Journal* 85 (1993): 609-618.

Khalil, Mounir. "Document Delivery: A Better Option?" *Library Journal* 118 no.2 (February 1,1993): 43-47.

Kibirige, Harry M. "Electronic full text Information Delivery: A Study of the U.S. Infrastructure." In Martha Williams ed. *Proceedings of the National Online Meeting* Medford, NJ: Learned Information, Inc., 1994.

Leach, Ronald G. and Judith E.Tribe. "Electronic Document Delivery: New Options for Libraries." *Journal of Academic Librarianship.* 18 no.6 (1993): 359-364.

Mitchell, Eleanor and Sheila A. Walters. *Document Delivery Services: Issues and Answers.* Medford, NJ: Learned Information, Inc., 1995.

Pullinger, David J. "Learning From Putting Electronic Journals on SuperJANET: the SuperJournal Project." *Interlending & Document Supply.* 23 no.1 (1994): 20-27.

Willis, Katherine et al. "TULIP—The University Licensing Program: Experience at the University of Michigan." *Serials Review* (Fall 1994): 39-47.

9. International Interfaces and Data Flow in Cyberspace

Information professionals who are faced with providing information to clients on demand can no longer depend solely on local, regional, or even national resources. Budgetary constraints have prohibited steady growth of in-house institutional resources at a time when user demands are soaring. Developments in cyberspace are progressively providing remedies to local information resource shortcomings. For organizations with the requisite internal information infrastructure, cybersources may provide a level of relief depending on local information needs.

A number of factors have come into play to make access to cybersources feasible and mandatory. First, internetworking at the national and international levels has provided the necessary connectivity. Second, free cyberspace information resources with downloadable segments go a long way towards satisfying client information needs. Third, client demand for information has become more diversified due to their increased exposure to the Internet. Ideally, information must be provided instantly regardless of source. Other than driving the information specialist almost crazy, these demands are taxing his or her information search capabilities to the limit. Fourth, there is a peculiar but welcome kind of camaraderie and collegial attitude that has been generated by the Internet. The amount of genuine helpful hints that one gets in cyberspace as part of this collegiality is a very pleasant surprise. Information professionals with tons of client problems (for example OPAC security issues or Internet access) get a chance to share them with other colleagues in similar situations. This level of resource sharing has made the existence of the Internet worthwhile indeed. Finally, hardware and software technology enables the information specialist to provide information to the client on demand using external national or international resources.

At the international level, there is a close linkage between electronic full text information delivery and networking and connectivity in general. International telecomunications networks facilitate information delivery. Many of the databases discussed in Chapter Eight were developed in different countries. Whether they provide citations or full text information, accessing and downloading some data from them depend on interfaces, standards, and protocols, provided on international networks. This chapter attempts to explore the major issues associated with information exchange on international information "Nets."

GLOBAL ELECTRONIC FULL TEXT INFORMATION DELIVERY

The Rationale

Within the U.S. information environment, attaining and maintaining super power status brings with it the need for tremendous aggregate international information. Usable information on demand requires a lot of full text information as an ingredient. Many of the most effective delivery systems that support the U.S. level of information consumption are electronic. Government, trade, industry, science, technology, and entertainment all thrive on international full text information. For example, world financial markets, especially stock markets in Japan, Britain, Hong Kong, and South Africa, affect the U.S. economy. Similarly, governmental change in key trade partners affects U.S. trade. This, in turn, affects production of goods and services, and eventually U.S. jobs and household incomes. While this may also be true for smaller countries, U.S. society consumes more international information than other nations of the world. In other words, the U.S. society flourishes on international information networks and the services they provide.

Second, within the U.S. economy, multinational corporations (MNCs), like AT&T, IBM, Exxon, Dupont, Firestone, and General Motors, have an insatiable need for foreign information due to their overseas subsidiaries. In addition, some small "niche" companies use foreign outposts as part of their business chain. By establishing relatively cheap overseas operations, they realize economies of scale and at the same time minimize overall production costs. Electronic technology, based companies, department stores, and drug manufacturing corporations, to mention a few, are regularly using these methods of generating products for an international consumer market. Full text information exchange between the headquarters and the subsidiaries, branches, or franchises, establishes the lifeblood links between the "home" office and overseas operations. Hitherto, much of the information traffic requiring speedy delivery has been characterized by hard copy. In order to beat the clock, delivery has been carried on by overnight, two-day, or some kind of express mail service. Companies like Federal Express, Emery Worldwide Company, and DHL Worldwide Express have serviced the hard copy equivalent of modern electronic documents for years. The use of electronic full text delivery is faster, more efficient, and cost-effective. Corporations that have adopted it for most of their operations have a competitive edge as the information plugged into their decision support systems is generated faster.

Third, the rest of the developed world that has large MNCs comparable to the U.S. has very similar full text delivery needs and the competing arena is the same. Britain has its Unilever, ICL, and British Petroleum which have overseas establishments. Within Germany, Siemens and Mercedes Benz play a similar role in the arena. Japan's Nippon, Toyota, and Mitsubishi likewise face the same issues. Finally, Italy's Agip and Sweden's Ericsson Information Systems also share the same competitive environment. This sampling of MNCs illustrates the need for electronic full text delivery of information for them to be fully competitive.

Finally, less developed countries in Africa, Asia, and Latin America are laying their own telecommunications foundations and networks. Nigeria, Botswana, and South Africa reside in this category. While many have rudimentary infrastructures, it will take some time before they fully participate in the electronic full text delivery environment. Part of the problem is the lack of a reliable electric supply on which the basic technologies are based. Fax has taken off very well in these regions and if it may be regarded as the precursor to full text information delivery, less developed countries may not be as far behind as some analysts seem to think.

Categories of International Full Text Information Delivery

Four major categories of electronic information delivery can be identified at the international level: intra-corporate delivery; inter-corporate delivery; user-to-user delivery; and corporate-to-user delivery. A variety of implementations has been used for these four categories, depending on the country or institution in question.

Intra-Corporate Flow

Intra-corporate delivery has always included a company's confidential information. During the hey-day of expedited mail delivery, sealed bags were delivered by express mail. For corporations which have established the infrastructure, electronic full text information delivery is a daily occurrence. Real-time transactions that monitor trade accounts, direct personnel and production actions, and general policy directives are carried out between subsidiaries and the home office. Competitive advantage is gained as production and distribution schedules can be co-ordinated at several global facilities without the deployment of supervisory personnel to and fro.

Manufacturing corporations often assemble different segments of a product in different parts of the world. The finished product is then assembled in plants near the markets. For example, an automobile manufacturing company targeting an East African market can have parts made in Singapore or Hong Kong and have the final product assembled in Kenya. The competitive edge gained by the use of electronic transborder data flow has been pointed out in studies and investigations carried out by the Organization for Economic Cooperation and Development (OECD) and the United Nations Commission on Transnational Corporations.

With regard to services covered, management of financial resources formed approximately 43 percent of the total transborder data flow traffic in a study conducted by OECD. Strategic information on marketing management and product distribution was a close second. With the increasing tendency to establish manufacturing plants in developing countries, partly because of cheap labor, graphics designs are becoming an important component for transborder data flow material. Projects that require computer-aided design (CAD), computer-aided manufacturing (CAM), and computer-aided engineering (CAE) are multiplying in number as a result of these trends. Designs developed in one country are electronically transmitted instantly to another to facilitate production on demand.

Intra-corporate transborder data flows can be ingeniously used to forge new business strategies in a given corporation. Corporate information centers at both the headquarters and in the overseas operations can coordinate information in collaboration with the management information systems personnel (MIS). Working within the copyright provisions of the countries covered, graphic designs, general policy, and marketing directives may be supplemented by technical papers screened from relevant literature. All this may be digitized at a central location or covered under distributed processing within the organizational network. In the latter case, more than one center may be used to develop digital versions of the documents to be shared in the organization.

Inter-Corporate Flow

Inter-corporate electronic information delivery is the second category that occurs at the transborder data flow level. It may take various forms. First, there are day-to-day business transactions. These may include electronic product orders; electronic fund transfers (especially among financial institutions), and travel arrangements (transport and hotel bookings). Second, there is the sale of information services and products by companies like Knight-Ridder and Dow Jones. Third, there is the exchange of general information about a given company provided on the Internet through a locally developed gopher, home page, or electronic bulletin board. Both foreign and U.S. companies have developed these services for other companies that are potential customers for their products. Such developments reduce the need for mass mailing general product catalogs on an international level, which is even more costly than local mailing.

User-to-User Flow

User to user transborder data flows on a personal level are becoming increasingly significant on the Internet and other networks. A large portion of users are individuals sending e-mail to each other, and exchanging data files of all types using the file transfer protocols on the international information "Nets." This type of connection on a personal level is gradually reducing the need for international telephone calls, travel, and some express mail deliveries.

Corporate-to-User Flows

Finally, the corporate-to-user mode of transborder data flow is happening, although at a relatively low level. Corporate gophers, home pages, and bulletin boards have sections meant for individual consumers rather than other corporations. International hotel chains have such services featuring products for individual customers. Electronic fund transfers, for individuals rather than institutional transfer, form a significant portion of the personal banking services that pay for international services and products. Hardware and software vendors also use this method to market products and support customers.

The Enabling Technology

Chapters Two, Three and Four discussed the technological foundation for international transborder data flow. The major intercontinental links for full text information delivery are made by satellite and transocean fiber optic cabling supplemented by regular national networks. Figure 9.1 depicts the main links in the northern hemisphere, which has the densest electronic intercontinental transborder data flow zone.

Within the U.S. two giant satellite organizations—COMSAT and INTELSAT—initially controlled all satellite communications. The Communications Satellite Corporation (COMSAT) was created by an Act of Congress in 1962. Its main objective was to provide ground facilities to access INTELSAT satellites. The International Telecommunications Satellite Organization (INTELSAT) was established by treaty in 1964 between 19 member nations. Its main objective was to launch and operate an extensive collection of satellites for international communications services. Membership has since increased to over 100 countries. Between them, the two organizations control much of the satellite communications between the U.S. and the rest of the world. They lease monthly rental services to both governments and corporations for civilian activity. Since the mid-1980s, the U.S. government has allowed private organizations to launch and manage satellites. A number of corporate giants like GTE Spacenet Corporation, Hughes Communications Inc., and GE American Communications Inc., have their own proprietary communications satellites. In addition, Department of Defense satellites are dedicated to defense activity, mainly surveillance. Some full text information delivery is transmitted via satellite.

The fiber optic based intercontinental trans-ocean cables (see figure 9.1) have been in existence for more than two decades.

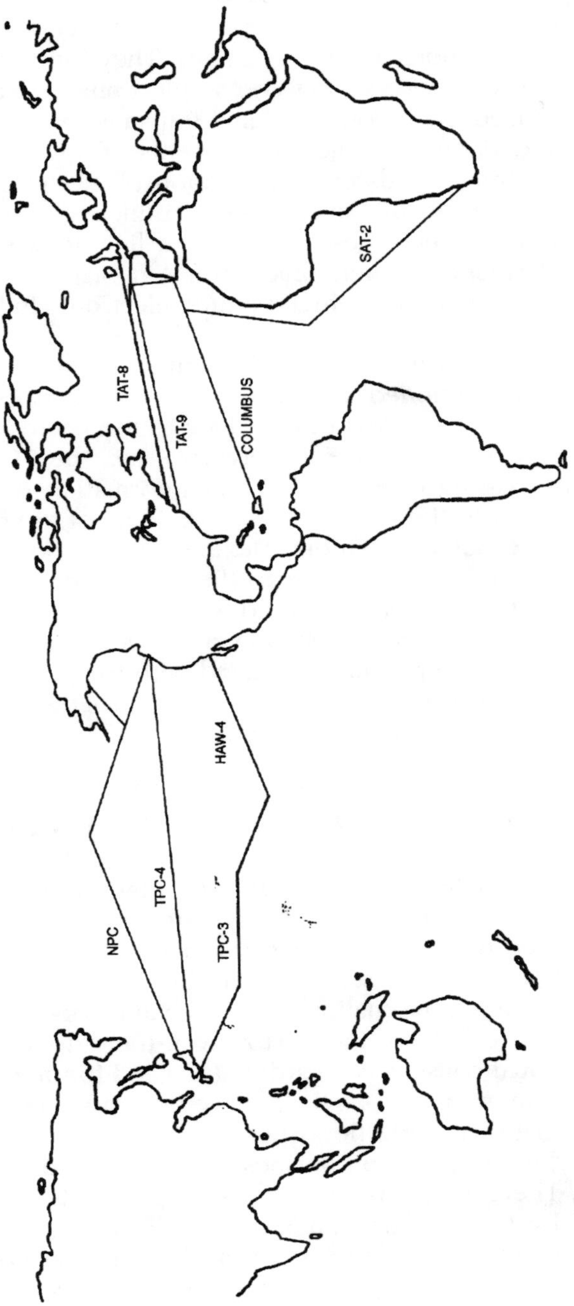

Figure 9.1 Transocean Underwater Fiber Optic Cables

They are owned by multinational corporations which lease usage time to other corporations and governments. They form the broadband links between national data communications networks and provide high-speed, high-volume signal transmission. Such links have not only defined the major components of the information superhighways, they are also the foundation of the future global village as they facilitate multimedia intercontinental communication. TAT-8 is one of the oldest connections, linking Western Europe to North America. Others link North America to Japan and the Pacific islands, permitting high-volume electronic full text information delivery.

As part of the enabling technology, computer hardware and software have been mentioned several times in the texts especially regarding the PC as an access device. Due to the decrease in the cost of the basic access device to networks, i.e., the PC, and its increase in processing capacity, full text information delivery is slowly inching out to the rest of the world. Some of the developing countries like Botswana, India, Mexico, Nigeria, South Africa, and Venezuela, are beginning to reap the benefits of transborder data flow using the Internet. This trend would have been curtailed if the main processing and access power had remained with the mini or mainframe computers. Associated with the growth of the PC has been the software sophistication that is progressively making the computer a less intimidating tool.

INTERNATIONAL STANDARDS AND PROTOCOLS

Standards that facilitate transborder data flow are in a state of flux as the international superhighways are unfolding. Chapter Three highlighted most of the issues associated with connectivity protocols and standards. If standards are needed at the national level, adherence at the international level is mandatory since there is no jurisdiction to enforce them. However, there appears to be a tacit international consensus regarding the need for international trade. Data communications standards have been adopted to make sure that international trade flourishes. Like elsewhere in the industrial sector, the information business has two broad categories of standards. There are formal standards, or protocols, which are promulgated by the various standardizing agencies, and the de facto standards which are imposed on a given industry or set of industries by a powerful organization or set of organizations.

TCP/IP is a de facto international standard imposed on the international community by U.S. technological power. It was developed under the aegis of the U.S. Defense Department for the ARPANET, and was initially used to link disparate computers and networks whose common denominator was contract research for the U.S. Defense Department. TCP/IP worked so well in the ARPANET environment that when the network was adopted by the National Science Foundation as NSFNET, it became an integral part of the new network which was dedicated to general research rather than defense research. Instead of developing another set of network standards, the rest of the world has adopted the TCP/IP suite and other standards are now developed around it. The three main elements of the Internet (File transfer protocol, Telnet, and e-mail) were developed under this suite and provide much of the formal basis for electronic full text information delivery on the Internet.

Formal standards for effecting transborder data flow are developed by national and international agencies, mainly the International Telephone and Telegraph Consultative Committee (CCITT) and the International Organization for Standardization (ISO). Between them, they have recommended standards that have been adopted by national standardizing committees.

CCITT is a French abbreviation for International Telephone and Telegraph Consultative Committee. It is a member of the International Telecommunications Union, a specialized agency of the United Nations. Although it was renamed the Telecommunication Standardization Bureau (TSB) in 1993, many of its standards have retained the CCITT title. Governments, like the U.S. (represented by the State Department), and some corporations, like AT&T and GE, are participating members. CCITT recommendations are treated as international standards. For example, in the E.500, E.600, and E.700 series of recommendations, CCITT has proposed guidelines that affect data communication performance parameters and end-to-end connections for international networks. These recommendations affect throughput and the quality of full text information delivery. CCITT recommended standard X.25 for packet networks. It defines the procedure for exchanging data between user devices (DTE) and a packet network node (DCE). Its formal title is "Interface Between Data Terminal Equipment and Data Circuit Terminating Equipment for Terminals Operating in the Packet Mode on Public Data Networks." Many of the networks that form the Internet are packet switched and the so called "packets" do carry full text information among other things.

Another important international standardizing agency is the International Organization for Standardization (ISO). As discussed in Chapter Three, it is best known for the Open Systems Interconnect Reference Model usually referred to as the ISO/OSI. This model is used for developing standards and protocols for providing connectivity to information networks. ISO is a voluntary organization composed of national and regional standardizing agencies, such as the American National Standards Institution (ANSI), the British Standards Institution (BSI), the European Telecommunications Standards Institute (ETSI), and the like.

Several specific standards and protocols have been suggested by ISO committees. Some networks use its file transfer protocol called File Transfer, Access and Management (FTAM). Using this protocol, a user may transfer either the whole file or parts of a file. It contrasts with the Internet's FTP, which transfers only complete files. Two protocols of special relevance to information professionals are the Search and Retrieve (SR) and the Inter Library Loan (ILL) developed by ISO Technical Committee 46. The sets of protocols for SR and ILL permit varying degrees of exchanging full text information among collaborating institutions.

DATA FLOW PROBLEMS

Over the last two decades, transborder data flow (TBDF) has generated problems and issues that have included copyright security, national sovereignty, and free flow vs. regulated information. At the international level, wealthy countries are pitted against poor ones. On an individual level, some critics have decried the gradual computerization and privatization of information resources in cyberspace at the expense of individual access.

General Issues

There are issues that affect all players in the TBDF game regardless of economic development. First, the problem of intellectual property characterized by copyright is one of the most sticky issues. International information networks have introduced a laissez-faire attitude among network users. It is virtually impossible for owners of intellectual works to detect copyright violations among users who meet on user interest groups or exchange professional information via e-mail. Nothing other than personal integrity will hinder individuals to copy, scan, or otherwise reproduce pieces of work. Computer software has been particularly vulnerable. Given the size of the Internet, the variety of users, recent cases of hackers, and computer virus developers, it is very unlikely that fraud, malice, and irresponsible behavior will be avoided.

One of the paradoxes of cyberspace is that even when a given country has instituted a copyright law, rights of foreigners may not be protected by it. A number of international conventions have passed guidelines affecting copyright: the Berne convention of 1886, the Rome Convention of 1961, the Geneva Convention of 1971, and the Brussels Convention of 1974 were all aimed at grappling with the problem. Countries which are signatories to agreements drafted at these conventions recognize foreigners' rights, but there are many countries which have not ratified the agreements. Another international device for protecting intellectual property has been the General Agreement on Tariffs and Trade (GATT). The Uruguay Round of GATT, signed on December 15, 1993, emphasized intellectual property rights, but again many countries are not GATT members.

Second, accountability and liability for information provided on international data communications networks poses another problem. International firms or individuals purchase and download factual electronic full text information to be used as part of their decision support systems. If information received in this manner is inaccurate, and a company loses money, the information vendor should be liable. Within the U.S., information processing companies have been sued for providing faulty information. Dun & Bradstreet vs. Greenmoss Builders Inc. [472 U.S. 749 (1985)] is the best known case involving an information processing company providing faulty information. Greenmoss sued Dun & Bradstreet for erroneously reporting that the firm had gone bankrupt. In actual fact, it was not the firm but one of its employees who had filed for bankruptcy. There is no equivalent conflict resolution arrangement at an international level. Consequently, several information consumers will have no recourse to recover damages caused by malpractice or negligence on the part of full text information providers.

Third, privacy of personal data is an issue that has not been resolved. A tremendous amount of personal data in full text electronic form is stored on computers owned by multinational corporations and other international organizations. This is vulnerable to abuse. The main concern is that foreign storage may facilitate foreign governments' access to sensitive data about countries where the information is collected. There are national laws that affect privacy, many of which are based on the U.S. 1974 Privacy Act. However, these laws are not standardized across national borders. The European Community (EC) has published a draft directive proposing the harmonization of protection between transmitting and receiving countries so that data is protected all along. An equivalent universal provision is not yet available and personal data cannot be examined for accuracy. A relatively common practice is for countries to prohibit transmission of electronic full text personal data to nations where privacy is lax or non-existent.

Fourth, national job dynamics are continually influenced by transborder electronic full text data flows. Satellite and cable networks can transmit full text information to manufacturing plants in parts of the world with cheap labor. Developing countries with stable political regimes have benefited from such plants. On the other hand, a sizable chunk of the job market in developed countries with high labor costs has disappeared. In addition, many developing countries are bypassed in this new international outsourcing practice.

Finally, culture domination has been cited in much of the literature as one of the problems caused by electronic information delivery. Cyberspace based on international networks has come to mean transmitting and receiving full text multimedia information. Since much of it is from the North Atlantic area, symbolizing the former colonial powers, a number of international critics have labeled it "electronic imperialism." Many countries are sensitive to the one way cultural influence that ignores local culture.

Transborder Data Flow Among Industrialized Countries

In spite of the competition for international markets, advanced countries of the North Atlantic welcome transborder data flow. This is also true of Japan and Pacific Rim countries. Estimates of the information sector as part of the GNP in industrialized countries is put at approximately 80 percent by the year 2000. As indicated in earlier sections, MNCs in developed countries flourish on information products and services. The European Community (EC) promotes the concept of Open Network Provision that encourages all EC members to develop connectable telecommunications services.

The EC

There are several programs that are being developed to help the EC manage the information age. Four are particularly interesting:

- *Research for Advanced Communication in Europe* (RACE) deals with the development of satellite and mobile communications to facilitate smooth flow of full text information on broadband and narrow bands.
- *Information Market Policy Actions* (IMPACT) monitors the information services market to stimulate competition among European firms.
- European Strategic Programme for Research and Development in Information Technology (ESPIRIT) deals with the development of standards for the information processing arena especially in Europe.
- Legal Advisory Board is the EC arm that makes proposals for circumventing legal barriers to transborder data flow.

The North-South Controversy

Triggered by a lack of resources to harness the information explosion, developing countries have raised objections to the dominant role of northern hemisphere countries. The southern hemisphere, containing much of Africa, Latin America, Asia, and the Pacific islands, has become the information age underdog. It has perpetually been on the receiving end for almost all information services and products (see figure 9.2). Very few products, for example databases, are developed in these countries. This state of affairs prompted the call for a New World Information and Communication Order (NWICO). It was literally the child of the MacBride Report of 1980, which in turn was sponsored by the United Nations Educational and Cultural organization (UNESCO). Prepared by several international contributors, the McBride Report highlighted several complaints. They are paraphrased as follows:

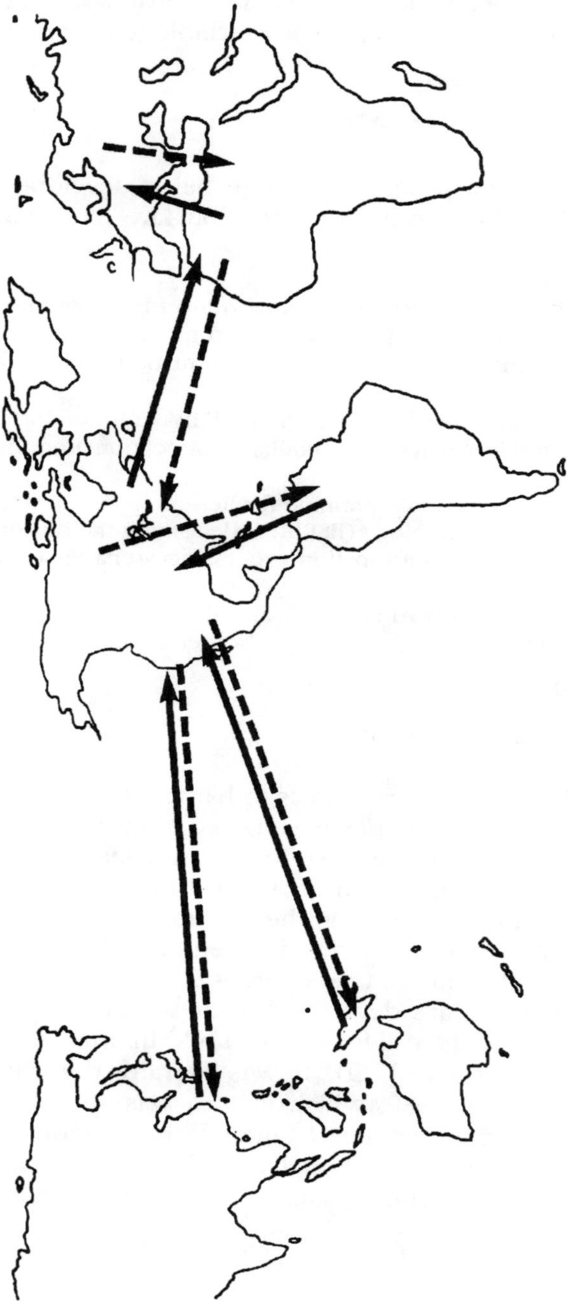

Figure 9.2 North/South Data Flow

1. Monopolization in production of information products and services by the North.
2. Misrepresentation of information on developing countries and a lack of substantive accurate information.
3. Perpetuation of cultural imperialism—manifest in advertising and exports of information products and services to developing countries.
4. Transmission of information that is skewed toward the developed countries and thus not relevant for developing countries.

While many of the complaints still persist, information professionals in developing countries realize the need for timely relevant information for economic growth. With help from industrialized countries, regional organizations have been formed to coordinate the development of regional information services. The International Development and Research Center (IDRC), a Canadian organization, has sponsored the Pan African Development Information System (PADIS), the Latin American Information Network (INFOLAN), and the Caribbean Information System for Economic and Social Planning (CARISPLAN).

In 1990, UNESCO published a handbook for developing national policies. It highlighted five major elements in national policy development: development of information resources; access to information; promotion of use of information; coordination of national information activities; and participation in international information activities. Among the developing countries with written national information policies are: Argentina, Brazil, India, Malaysia, Mexico, and Singapore. Such policies make up the framework for nationwide electronic full text information delivery systems.

References

Adeyemi, Nat M. "Issues in the Provision of Information to Developing Countries." *African Journal of Library, Archive and Information Science* 1 (April 1991): 1–8.

Bar, Fernando L. "Can Third World Countries Become Information Societies?" *International Journal of Micrographics and Optical Technology* 10 (1992): 71–73.

Bender, David R. "A Strategy for International Information Policy." *Libri* 43 no.3 (1993): 210–231.

Bing, Jon. Reflections on EC Data Policy. *Transnational Data and Communications Report.* (May 1990): 15–17.

Black, Uyless. *Computer Networks: Protocols, Standards, and Interfaces.* Englewood Cliffs, NJ: Prentice-Hall, 1987.

Bollinger, William A. "Information Access Policies in the 1990s: National and International Concerns." In Martha Williams ed. *Proceedings of the Eleventh National Online Meeting.* Medford, NJ: Learned Information, 1990.

Buga, Wladyslaw J. "AT&T International Services Support Architecture Overview." In *IEEE 1992 Network Operations and Management Symposium.* New York: IEEE, 1992.

Callendar, Mike. "Future Public Land Mobile Telecommunication Systems." *Canadian Journal of Electrical and Computing Engineering* 17 no. 4 (1992): 152–155.

Figueredo, Nice. "Information as a Tool for Development." *The International Information and Library Review* 24 (1992): 189–201.

Gosztony, Geza. "CCITT Work in Teletraffic Engineering." *IEEE Journal on Selected Areas in Communications* 9 no. 2 (1991): 131–134.

Holm, Liv A. "Connectivity, and Protocols—The Technical Side: OSI, and TCP/IP, FTP, TELNET, SR, ILL, Update." *Ifla Journal* 20 no. 2 (1994): 158–170.

Howe, Eric. "The United Kingdom's Data Protection Act." *Government Information Quarterly* 8 no.4 (1991): 345–357.

Kluepfel, Henry M. "Securing A Global Village and Its Resources." *IEEE Communications Magazine* (September 1994): 82–89.

Kraemer, Kenneth L. et al. "Economic Development, Government Policy, and the Diffusion of Computing in Asian-Pacific Countries." *Public Administration Review* 52 no.2 (March 1992): 146–156.

MacBride, S. *Many Voices, One World: Towards a New More Just and More Efficient World Information and Communication Order.* New York: Unipub, 1980.

Mantviloff, Victor. *National Information Policies: A Handbook on the Formulation, Approval, Implementation and Operation of a National Policy on Information.* Paris: UNESCO, 1990.

Masmoudi, M. "The New Information Order." *Journal of Communication* 29 (1979): 172–185.

Organization for Economic Co-operation and Development. *Information Networks and New Technologies: Opportunities and Policy Implications for the 1990s. Information Computer Communications Policy No 3.* Paris: OECD, 1992.

Perez, Victor L. "Developing Informatics Policies." *Information and Management* 23 (1992): 7–14.

Plesser, Ronald L. and Emilio W. Cividanes. "EC Personal Data Privacy: US Concerns." *Transnational Data and Communications Report.* (November 1990): 19–20.

Preer, Jean. “Special Ethics for Special Libraries.” *Special Libraries* 82 no.1 (Winter 1991): 12–19.

Sauvant, Karl P. *International Transactions in Services: The Politics of Transborder Data Flows*. Bolder, Co: Westview Press, 1986.

Schoonmaker, Sara. “Trading Online: Information Flows in Advanced Capitalism.” *The Information Society*. (1993): 39–49.

Sharpe, Diana M. “Asia-Pacific TDF Issues.” *Transnational Data and Communication*. (May 1990): 220–24.

Swain, Leigh and Paula Tallim. “The Interlibrary Loan (ILL) Protocol: Progress and Project.” *IFLA Journal* 8 no.4 (1992): 324–332.

Wang, P. “Information Systems Solutions For Transborder Data Flows for Multinational Companies.” *International Journal of Information Management*. 13 (1993): 29–40.

10. The Human Side of Information Access

As an economic development tool, information is becoming invaluable for national economies. Ongoing technological development makes full text delivery the focus of the worldwide information environment. By the year 2020 most industrialized countries will derive more than 80 percent of their GNP from the information sector. At the individual person's level, information is becoming vital for survival. It is not sufficient just to be able to read. To be a responsible citizen in a participative democracy like the U.S., one must be able to access, scan, acquire, and assimilate large amounts of information. As we approach the twenty first century, information will be more available in digital form than in hard copy. Ideally digitization should make information faster to generate, disseminate, and acquire. Alas, information is not becoming more affordable for the average individual, nor is it for small corporations.

Politicking and theorizing aside, one would have thought that at a pragmatic level, when all the bells and whistles of the information highway are in place, there will be a marked improvement in the human condition. While the structure being established will undoubtedly be a vital economic investment, it will most likely benefit the haves more than the have-nots. It will not be easily accessible to that person in the street who is responsible for much of the basic capital used to finance it. A few questions must be asked at the outset to portray the human side of information access.

Assuming that information is invaluable for individual and institutional survival, should access not be universalized? If so who should foot the bill for developing the infrastructure? Should online information be subsidized for those who cannot pay? What is the relevance of the concept of information democracy? Some information centers already have access to the Internet. Should access be free? Should access to the networks be regulated? Are our networks safe? What methods can we use to secure them? How can we stop hacking and virus infection? A lot of personal confidential information is being gulped by databanks in government and private organizations. Selling or exchanging mailing lists among commercial firms is a common transaction. However, organizations do not often reveal to the customers included on the lists what kind of information about them, beyond the postal address has been peddled. Is individual privacy protected when sensitive personal lists or other data are exchanged or sold among organizations?

There are no easy answers to these seemingly simple questions. There are three basic alternatives to looking at the aggregate issues raised by them. Within the U.S. context, there is the proposition that government, whether federal, state, or local, should take over. This is overly simplistic and unrealistic given the current public opinion trends that stress downsizing government. The task at hand is so enormous that government alone cannot be expected to accomplish it. On the other hand, critics that dismiss the "Uncle Sam" approach tend to forget that "Uncle Sam" is in effect the concept of the collective effort of a polity. In its intrinsic application, the concept calls for a concerted nationwide effort to tackle apparently insurmountable tasks. This has worked in times of war and when building interstate highways. It has also worked in establishing the prototype NSFNET backbone that launched the U.S. component of the Internet.

The second, more popular approach is to leave it to the market forces, both for establishing and pricing of the National Information Infrastructure (NII) based resources that provide full text access to information. This typical economic argument is buttressed by the capitalist premise. Entrepreneurs develop services and products. Truly useful products that the market can sustain will thrive and those that are not on demand will be weeded out. The price equilibrium theory will ultimately regulate the market price such that if the price is too high, the product will not sell. In other words, the consumer market will dictate an optimum price acceptable both to the vendor and buyer. With open competition, there will be so many products that prices will come down to affordable levels. However, studies of poverty and wealth in the United States indicate that the gap between the wealthiest and the poorest American is widening. The affordability of information products could be relatively high, shutting off a sizable portion of the population from electronic based information.

A compromise approach might be a hybrid arrangement

whereby government support is given for some of the projects and services which may not be profitable, but are still needed to improve the human condition. At the same time entrepreneurship would be encouraged to develop products and services for the consumer market. Whether government at the various levels wants it or not, by the very nature of the U.S. society, privately developed information products and services will be very significant components of the traffic on the information superhighways. The need for government seed money to bolster expensive experimental projects will also continue.

This chapter explores the human factors in full text information access. It does not purport to provide panaceas, but rather to highlight pertinent challenges that stand in the way of achieving equitable full text information delivery. A basic assumption made throughout the chapter is that however constituted, the NII aims to improve the human condition.

THE CHANGING VENDOR ENVIRONMENT

It is perhaps expected that with improved information transmission, information products and services will equally be diversified. Some parts of this text have mentioned new information services and products generated around the edifice of the "freenets." What we are witnessing is the development of completely new products that have never been on the market before. The basic question is how does this affect the human side of information access? Does new necessarily mean better services for the end-user or better profit opportunities for the vendor?

The information marketplace may be broadly defined to include vendors of hardware, software, telecommunications services, databanks, and cyberspace products of the future. In other words, it may be categorized into vendors of processing devices and associated search engines; vendors of telecommunications devices and accompanying processing engines; and vendors of information content used in responding to human activity. Our emphasis here will be on vendors of information content, sometimes referred to as online information services. Such a focus brings us closer to vendors that facilitate full text information delivery.

Initially, most online services were derived from the need for fast access to information also published in hard copy. Since the late sixties, Dialog Information Services (now Knight-Ridder), Bibliographical Retrieval Services (BRS), and I.P. Sharp, to mention a few, have had gigantic databases composed of mainly machine readable versions of indexes and abstracts such as Chemical Abstracts, Psychological Abstracts, Index Medicus, Applied Science and Technology Index, and the like. The earliest online services offered only bibliographic information composed of abstracts and indexes. To get the substantive full text of the needed article or report, individuals had to request much of the identified material through interlibrary loan. With the sophistication of hardware and software came full text delivery services in the late 1970s, the most comprehensive being in the legal realm, as exemplified by Westlaw and LEXIS. Other online vendors like Dialog added full-text databases to their databank collections. These "full-text" databases lacked graphics (charts, tables, and pictures) that were included in the original published documents. Most of the early online services targeted government, business, law, research, and academic institutions. They had virtually no material for entertainment or other general information.

The 1980s and 1990s ushered in online services that covered general purpose type information. New vendors included: Prodigy, Compuserve, The Source, American Online, and the like. Among the services the new vendors provided were, travel guides, telemarketing for businesses, e-mail and, access to the Internet. In addition, they provided a gateway facility to other services, so that Compuserve subscribers may connect to Knight-Ridder information services through Compuserve if they choose to do so on condition that they have Knight-Ridder access numbers.

A third category of online service that aims at taking advantage of the information superhighway was launched by AT&T and Microsoft. AT&T's Interchange and the Microsoft Network are relatively new services that provide entrepreneurs who have developed information products an opportunity for marketing them via cyberspace. The main difference from earlier online offers is that Interchange and MSNetwork provide environments on the information superhighway rather than databases per se. They provide service mainly to companies that have a database or databases (a product with information content) that they want to put on the superhighway for sale.

Vendors of full text information are thus changing with regard to products and services provided. Several services in cyberspace allow users to communicate relatively cheaply, exchange documents, shop and have goods and services delivered to their premises. These services have definitely eased the way individuals obtain conventional services that might have otherwise required trips to sales offices and waits in long lines. On the other hand, in order to use the services, one must have access to a computer either at home or at work. Since computers at work are ordinarily used for work-related activities, the individual who benefits most is one with a home computer. Recent studies of home computer ownership put it at between 40–50 percent of U.S. households. A sizable number of home computers are not connected to any network, thus making the per capita number of users of these services quite small. Equity of information access is called into play and discussed in the next section.

THE HAVES, THE HAVE-NOTS, AND INFORMATION DEMOCRACY

The federal government's statement of intent about the National Information Infrastructure (NII) sounds impressive but can it deliver the implicit promise? *The National Information Infrastructure: Agenda for Action* document published in 1993 by the U.S. Department of Commerce stimulated very high hopes when it said:

> All Americans have a stake in the construction of an advanced National Information Infrastructure (NII), a seamless web of communications networks, computers, databases, and consumer electronics that will put vast amounts of information at users' fingertips. Development of the NII can help unleash an information revolution that will change forever the way people live, work, and interact with each other.

An effective NII, if it materializes, will indeed change the way lots of people in the U.S. interact. The Internet has already changed many peoples' ways of life, but so far it has been a fairly small fraction of the population. In spite of the progressive lowering of price for computer systems that has occurred over the last two decades, most Americans do not have easy access to computers that are connected to networks.

Ronald Doctor has suggested using the NII for empowerment of American citizens using the concept of information democracy. He defined information democracy as, " . . . a socio-political system in which all people are guaranteed meaningful opportunities to benefit from access to information resources." In this context, empowerment may have a number of connotations.

First, easy and equitable access to relevant politically oriented information empowers individuals to make informed decisions with respect to participative democracy. In fact, during an election year, federal and local governments should make full text online information readily available to the electorate. Such information should include summaries of the electoral process, new issues on ballots, and objective information about the candidates. For national candidates, biographical scoops can be obtained from available data banks, but the cost is high and the ordinary citizen's access is constrained by lack of access devices.

Second, easy access to full text information empowers a person to make informed decisions about job opportunities that may widen chances of employment. Public job banks with associated skills schedules can be created in cyberspace and easily searched and downloaded. While employment agencies might object to this type of intrusion into their business, a job-seeking individual will be empowered to make his or her own judgment. Currently listservs on the Internet and some newsgroups on networks advertise job openings, but access is limited.

Third, equitable access to personal communication on the network enhances self-esteem and may reduce anxiety and stress. National and international pen pals have to a certain extent resulted in offsetting loneliness. In addition, networks have developed a lot of socio-referent groups that serve as virtual communities for pure entertainment and light humor. Again these groups do to some extent resolve some personal human tensions. In other words, information access empowers individuals to become members of "virtual clubs."

Fourth, easy information access promotes both formal and informal education. Students can access databank information that helps to enhance their studies. Most college computer labs are overcrowded, especially in institutions which don't require that individual students purchase personal computers at the time of admission. Students who live off campus should be able to dial into educational databanks without incurring exorbitant telecommunications charges. Furthermore, within the K-12 educational environment NII should aim at connecting teachers, students, and parents in the educational enterprise. Some schools have established bulletin boards accessible via the Internet, where parents and teachers may discuss common educational concerns. With regard to informal education, preparation of guides to general resources is being done by voluntary and commercial organizations.

Easy access is inhibited by network access costs and the very basic requirement of possessing a PC and installing a network connection.

Creating and maintaining general purpose databases and establishing national and regional information agencies has been suggested by some analysts of the NII. At the national level, a federal agency to coordinate information generation and distribution has been suggested with various designations like "National Database Server Center" or the National Institute for Information Democracy (NIID). This agency would coordinate regional information agencies with a structure somewhat comparable to the National Institutes of Health. NIID would create and disseminate national information services. It would also act as a clearing house for regional agencies. This is a radical departure from the current government information set-up whereby federal agencies have machine readable online full text information databanks about their services and documents they publish. As already discussed in Chapter Seven, The Office of Management and Budget is currently overseeing the agency information policies to ascertain uniform practices. At the same time, the National Technical Information Service (NTIS) has a service called "Fedlink" that acts as a gateway to all other federal agencies.

As expected, strong opposition comes from members of the Information Industry Association as government becomes a direct competitor in providing information products and services. However, this is not a new development as a number of federal departments have developed online services that they have sold to end-users or other institutions. The National Library of Medicine has for decades developed and sold the Medline database. Similarly, the National Agricultural Library and the Department of Energy have had on the market, AGRICOLA and Energy Science and Technology, respectively.

As the NII develops, three possible approaches may emerge. First, information on government remains the function of the various agencies, coordinated by the Office of Management and Budget. Second, databases on professional activity (MEDLINE, AGRICOLA, DOE ENERGY) which have national and international coverage are developed, comparable to any other databases. Third, the government develops general purpose databases. While the actual strategy may not be that difficult to design, it appears to be on a collision course with the private information providers. At the same time, if the government does not provide such a service in the name of equitable access for the have-nots, none of the commercial vendors would service the information poor who incidentally are also the economic poor.

Effect of Copyright on Equitable Access to Information

Authors of hard copy and electronic products have justifiable concerns about piracy on information networks. Hard copy is not safe, in as much of it can be scanned and transmitted as electronic files. Many of the concerns from the producer's point of view have been covered in Chapter Nine. With regard to the information poor, copyright royalties add to the high cost of accessing electronic full text information. Copyright, which essentially establishes intellectual property rights, is thus in direct conflict with access rights for the information poor. Treatment of the information poor should be compared to the treatment of the poor or senior citizens who are sick. Medicare and Medicaid are society's responses to the medical problems of these two categories of citizens. Similarly, society must recognize individuals who need access to information in order to make a living, but who cannot afford it. *Subsidized access to electronic information must be one of U.S. society's responsibilities.*

In other words, a form of subsidy has to be worked out for the information poor, most likely through the Copyright Clearance Center (CCC) which already exists. Currently, many document delivery services, which are intensive users of copyrighted material, pay royalties through the CCC. Similarly, if the concept of a National Database Server Center (NDSC) and regional agencies is accepted, the economically disadvantaged would get electronic information from these organizations which would take care of copyright via the CCC. The point of access to the NDSC could be an OPAC from a library that agrees to be part of the system.

Equitable Access Regulation or Mediation?

It may be argued that after the 1996 Telecommunications Act, the issue of regulating full text information delivery is moot. By further deregulation of the telecommunications industry, the U.S. released the Baby Bells to participate in long distance services and develop information services. Likewise, cable companies may deliver other services other than regular cable. The government has over the years practiced both regulation and mediation of the telecommunications sector of the information industry. The complaints about AT&T that led to its divestiture in 1984 were mediated by the judiciary and the legislature while the executive branch and the states regulated state and interstate telecommunications services.

The 1996 Telecommunications Act does not absolve the government of its responsibility to continue both mediation and regulation. Unlike the closely regulated power utility companies the new arena created by the Act provides for minimal levels of regulation as rates are to be dictated by market forces. There is, however, no specific provision that addresses equity of access to the services to be provided. Every citizen will have access to information services and products according to their ability to pay. This leaves out of the loop a large amount of have-nots. A lot of services may be comparable to "pay TV" channels. Since many of the products and services are not yet developed, one would think that appropriate amendments would be devised as need arises.

TELECOMMUTING WITH FULL TEXT INFORMATION

Sophistication of the transmission media, software, and hardware, as well as increased computer literacy among users, have made working at home feasible. There are several definitions of telecommuting and the one used in this section is adapted from Mokhatarian and Sato: Telecommuting is defined as working at home or a location near home in lieu of the primary office, but with realtime telecommunication connection with the primary office (see figure 10.1). It is one of the prime applications that makes optimal use of the concept of full text information delivery. With appropriate equipment, an employee can send and receive from the main office, formal information that may have sound, text, and images. When given an opportunity to telecommute, many people have opted for it. In 1992, Charbuck and Young estimated that seven million Americans telecommuted to work. Given the flexibility and availability of the enabling technology, the trend is bound to continue or even accelerate as we approach the twenty-first century.

There are several reasons why employees and employers have willingly embraced the telecommuting bandwagon. A federal le-

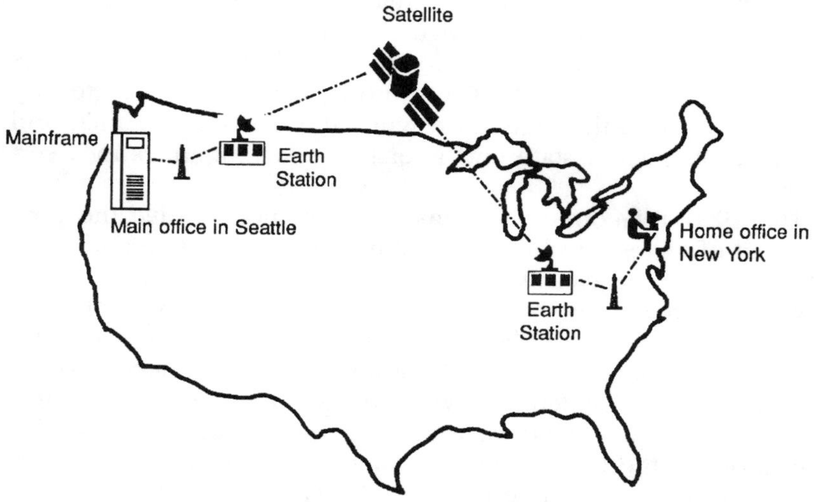

Figure 10.1 Telecommuting

gal requirement stipulates that urban areas with serious air pollution reduce automobile commuting by 25 percent. Los Angeles is one such metropolitan area which encourages car owners to use public transportation in order to cut down on pollution. This is done through employers, affecting organizations with more than 100 employees. In addition, telecommuting enables employers to cater to employees with disabilities by allowing them to work at home in compliance with the Americans With Disabilities Act (ADA). A third incentive for employers is the saving of office space for those who do not need desks and filing cabinets at the main office. Depending on the nature of the business, office space would be progressively reduced due to disk online storage of official documents. Such a trend would increase to facilitate telecommuting. Third, for employees who have small children and would rather work at home, their morale, and thus effectiveness, would be improved. This depends on whether the job will permit working around children. Fourth, again depending on the nature of the job, employees may not need to move from state to state to work at different offices. This would reduce moving expenses if provided by the institution.

On the negative side for the employer, monitoring performance might be difficult. Evaluation changes from how much time is spent on the job to assessing outcomes or results. Second, with remote access to full text data, employees may feel isolated and motivation may be more difficult to achieve as physical detachment may eliminate a sense of belonging. Third, the confidentiality of company information may be compromised when two friends work at home for competing firms. Finally, depending on the nature of work to be accomplished, start-up costs per employee may be high ($6,000–10,000).

With regard to employees, incentives include reduction of the commute time, which in turn reduces car maintenance and parking expenses. Second, since casual clothing is used at home, the cost of "business type" clothing will certainly be reduced. In addition to casual clothes, the employee controls the work environment (temperature, lighting, decor) according to personal interest. The total ambiance created by "doing your thing" might decrease personal tensions or stress and improve the individual levels of productivity.

The employee who telecommutes has a negative side, too, with regard to work-employee relations. First, personal isolation is real and e-mail or bulletin boards may not be substitutes for human personal interaction. Coffee, lunch breaks, and staff meetings provide the institutional bond which is broken by telecommuting. Second, depending on how telecommuters are classified, whether part-time or contract, they may lose fringe benefits that normally accrue to full-time personnel. This may vary from organization to organization and the employment contract mutually agreed on by the employee and the employer. Third, an off-site employee may be bypassed at promotion time unless the job allows star quality to be noticed, like sales people with very high sales volumes or software engineers who have developed state-of-the-art software packages. Finally, unless the telecommuter is assertive, he or she may not participate in informal problem solving sessions at the office. Informal chats or discussions may occur among professionals working on similar, but not necessarily identical projects. Coffee breaks sometimes serve as small scale brain storming or venting sessions during which solutions to difficult tasks are exchanged. Telecommuters normally miss such informal professional contacts.

TWO IMPORTANT PROBLEMS

Privacy and Full Text Information Delivery in Cyberspace

Since full text information delivery systems may provide the fullest records on individuals, personal privacy is a significant factor in their management. Privacy of personal records has not been dealt with satisfactorily on national or international information networks to date. This may be a reflection of the multi-jurisdiction nature of the networks in question. Interestingly, many of the developed countries have legislation restricting government intrusion or interception of private electronic messages. This may perhaps be due to the recognition of the fact that governments have the means to eavesdrop. Government monitoring of private information on networks may be legitimatized by court order. The onus is on the government to prove that such prying is needed for national security.

Within the U.S. the 1986 Electronic Communications Privacy Act protects cyberspace communications from government interception or interference of any sort. It is partly a legacy of the Watergate surveillance of the 1970s which helped pass the 1974 Privacy Act. Protection extends to in-transit as well as stored messages. Many western countries have somewhat similar legislation that bars the government from acquisition or distribution of private confidential information about individuals. A lot of electronic full text information is in non-government institutions like banking, healthcare, and insurance, which do not have similar restrictions.

With regard to individual electronic snooping on others' information, there are virtually no international or national legal restraints. With recent reports of break-ins into university, corporate, and government databanks, legal restrictions are needed at the national level. Access to sensitive personal or organizational information may result in sabotage, blackmail, or extortion.

Security and Vandalism in Full Text Delivery

A rather simplistic outlook to the world at large might be that if there were no malice, vindictiveness, ambition, curiosity and in sum "human vice," we would not have to worry about hacking, virus planting, and the like. Alas, human nature has all these characteristics: thus the need for securing full text information storage and transmission systems is paramount. Similarly, viruses that may incapacitate full text delivery systems will always be with us. Otherwise, we may take pride in our delivery systems when in fact the full text information either stored or delivered is adulterated by vandalism or becomes temporarily inaccessible due to virus infections.

Security of information systems ensures that information stored or transmitted is not tampered with by external unauthorized users. It is usually in the purview of systems designers or managers. The end-user simply gets whatever is provided. However, it is important to know some of the precautions that may minimize intrusion or virus attack.

The most secure system against outside invasion is either a stand-alone or part of a local area network (LAN) which is not connected to external networks through gateways or bridges. With respect to data access, there are programs on the market ("Fire Walls") that prohibit unauthorized entry into an electronic information system. In addition, encryption—scrambling data at either the storage or transmission level or at both levels—is another way of ensuring the security of electronic full text data. At the receiver's end, decryption before use is needed to make the data usable.

Computer viruses can render a full text information delivery system dysfunctional. By definition, a computer virus is a computer program which attaches itself to executable systems programs on the computer system. When executed, it may erase either data or systems programs. One of the precautions of avoiding viruses is the use of anti-virus software. This may be memory resident and would start when the system is turned on or can be on a separate disk which is used to clean the system.

References

Churbuck, David C. and Jeffrey S. Young. "The Virtual Workplace." *Forbes* 150 no. 12 (1992): 184–190.

Doctor, Ronald D. "Seeking Equity in the National Information Infrastructure." *Internet Research* 4 no. v3 (1994): 9–22.

Fleming, Maureen. "Microsoft Network and Interchange: Changing Online Metaphors." *Online* 19 no.2 (1995): 71–72.

Glory, Blaise. "Managing Information Resources in a Telecommuting Environment." *Special Libraries* 85 no. 1 (1994): 30–34.

Hallgren, Martyne M. "Funding an Internet Public Good: Definition and Example." *Computer Networks and ISDN Systems* 27 no. 3 (1994): 403–409.

Hernon, Peter. "Equity in Public Access to Government Information." *Government Information Quarterly* 10 no. 3 (1993): 301–304.

Kelly, Brian. "Becoming an Information Provider on the World Wide Web." *Computer Networks and ISDN Systems* 27 no. 3 (1994): 353–360.

Martinez, Michael E. "Computers and Social Equity: Views From the National Assessment of Educational Progress (NAEP)." In *Proceedings of The 55th ASIS Annual Meeting of The American Society for Information Science, Pittsburgh, PA. October 26–29.* Debrah Shaw ed. Medford, NJ: Learned Information, 1992.

McMullen, Melanie ed. *Networks 2000: Internet, Information Superhighway, Multimedia Networks and Beyond.* San Francisco: Miller Freeman Books, 1994.

Mokhtarian, Patricia L. and Koji Sato. "A Comparison of the Policy, Social, and Cultural Contexts for Telecommuting in Japan and the United States." *Social Science Computer Review* 12 no.4 (1994): 641–658.

Romei, Lura K. "Telecommuting: A Workstyle Revolution?" *Modern Office Technology* 37 no.5 (1991): 38–40.

Tonn, Bruce E. "Using the National Information Infrastructure for Social Science, Education, and Informed Decision Making." *Social Science Computer Review* 12 no.2 (1994): 166–182.

Turkheimer, Frank. "Privacy and the Internet: The Next Step." *Computer Networks and ISDN Systems* 27 no.3 (1994): 395–401.

Williams, Martha E. "The Internet: Implications for the Information Industry and Database Providers." *Online & CDROM Review* 18 no.3 (1994): 149–156.

Young, J.A. "The Advantages of Telecommuting." *Management Review* 80 no.7 (1991): 19–21.

11. Harnessing the Electronic Full Text Superflow in Cyberspace

We live in an era when access to information determines the destinies of individuals as well as organizations. Full text information that includes text, images, and sound facilitates decision making with what economists call "perfect information." Organizations need such information to be competitive in global markets. Similarly, achieving individuals have to use information-based tools for sustained excellence. Such tools may include: online databases, freenets, electronic bulletin boards, and web sites. Society has to take advantage of the information environment in which full text information delivery is progressively becoming the norm rather than the exception. This is particularly true in the advanced countries of Western Europe, North America, Japan, and the newly industrialized countries of the Pacific Rim.

This text has emphasized the importance of full text delivery systems as they affect national economies, organizational culture, and the individual's self actualization. It is desirable to harness the information around us so that we can use it on demand and to our advantage. The implementation of such a concept is at best a tentative proposition. We must understand the trends that lead to the realization of this concept. At the same time, we must be aware of the constraints—social, political, economical, and environmental—that are working counter to optimum realization of electronic information flow.

As we get ready to plunge into the twenty first century, computers and telecommunications networks must be understood for effective planning of management information systems (MIS) and information access centers. Discussion of the basic elements of networks and networking was done in the first two chapters of the book. This was done in order to underscore their underlying function in information delivery. It is true that in a technology-driven society, everything conceivable is in a dynamic state of flux and "change" becomes the constant. Keeping abreast of the changes can only be maintained by regular access to full text information available in cyberspace.

Eliminating the Archipelago Syndrome

I have also attempted to demonstrate the need for implementing the concepts of connectivity and interoperability in order to avoid the information archipelago syndrome, in which networks do not talk to each other. Networks are analogous to islands which are separated by water and thus the use of the archipelago metaphor. Networking constitutes both the salvation and the bane for information providers. They are salvation in that databases in one country can be accessed from another only when intermediary networks have connectivity, and their resident programs are interoperable. At the same time, they are a bane in that malicious damage and sabotage to systems has been caused by systems talking to each other. Computer virus infection, software pirating, and deliberate data contamination also comes with system connectivity. Given the resources, sensitive data must be on unnetworked computer systems, or if networked, on networks which do not have gateways to other networks.

Malicious hacking and virus implantation aside, organizations cannot avoid participation in the access and distribution of electronic full text information. I have demonstrated with examples how the use of public computer directories, or disks that can be accessed with anonymous FTP, will eventually save marketing departments millions in document preparation, printing, and postage. Instead of bulk mailing of catalogs, clients can access a company's public disk to obtain product information online. Orders can also be made online. In addition, instant access to corporate full text data by traveling personnel, like sales people and system maintenance staff, will be more effective as they serve the global markets. Using organizational (corporate) bulletin boards for selective advertising will also cut down on the cost of regular hard-copy mass market advertising, which is essentially a hit or miss exercise. In other words, organizations that utilize the facilities in cyberspace for strategic planning and repositioning of business enterprises will certainly be at a competitive advantage.

Information Dynamics in Re-engineering

Re-engineering, in both for-profit and non-profit organizations, becomes more critical as we approach the uncertain twenty-first century. Corporations must have accurate internal evaluations in order to rejuvenate themselves. Depending on the organization, the way internal information is managed will determine how fast it can respond to critical decision making on demand. Decision support systems must be in place for staff to quickly pull out relevant information to be used for analysis. The most effective information remains full text. Decision support systems must thus be redesigned to be able to provide it.

It is an absolute necessity for organizations to make informed decisions about their positions *vis-à-vis* their competitors. Environmental scanning is a concept used in business to refer to activities performed by an organization to fully appraise itself of external factors that affect its effectiveness. This is a vital procedure to be done prior to re-engineering. Much of the information about market share and performance are available in electronic form. The SEC's EDGER online program which provides company profiles, is one of the best examples of emerging electronic sources that facilitate the comparison of organizations. Full text information delivery is invaluable in environmental scanning as it provides part of the information used in generating the composite picture of the organization prior to, and after, the re-engineering phase.

Networking for Human values

As we become immersed in cyberspace, with its distinct cyberculture, we may get carried away and forget that information systems are designed for human usage as utilities to improve the human condition. A critical examination of some of the systems and their idiosyncracies makes one wonder whether they were designed for the designer's pleasure or for effective end-use. Reading some of the documentation that accompanies software reveals little consideration for the user. This may be because the end-user population is so diverse and novices as well as veterans are difficult to accommodate in a one-size-fits-all type of document. Many of the hard copy manuals, as well as online help, are written in unintelligible jargon which the end-user may not be able to comprehend. In addition, access to vital information in electronic form is an elitist exercise due to the cost and sophistication required to get it.

Within the U.S., the National Information Infrastructure (NII) seems to have promised more than the government can deliver.

The financial commitment that is needed to implement the program is just too much. Such good intentions were counteracted by the 1996 Telecommunications Act which allows for a laissez-faire market share approach. Such an approach has little regard for those individuals in society who cannot afford to pay to get on the information superhighway and reap the benefits of the NII. The haves often make the mistake of thinking that it is the fault of the have-nots that they lack the means to access the vital electronic information that they need for self-improvement and survival. This problem is becoming more serious as much of the information once available in hard copy is increasingly becoming available only in electronic form. Furthermore, an increasing amount of information is emerging which has never been and will never be available in any form but electronic. The haves must devise ways of ensuring a degree of equitable access to such information and avoid hoodwinking their critics by using the elitist adage, "What you do not know can't hurt you." After all, they might argue that the have-nots are not aware of full text information delivery systems. Is "equitable" fashionable in cyberculture?

Unanswered Questions

Writing about information technology tends to invite criticism about the many issues mentioned and questions raised with no ready answers provided. For example, Do we know the information needs of the poor? How about families in rural areas? Who will provide the funding for access links for the unserved? In some cases the issues seem too commonplace to have even been mentioned in a book. Alas, we have a way of ignoring the obvious until it strikes us right in the face. If this text raises the reader's consciousness of the importance of full-text information delivery systems and their significance in ordinary peoples lives, it will have served its purpose. As information professionals, as well as users, we need to pool our intellectual resources to make information a truly human resource to be shared by all for optimum self development. This may be shown in work-related activities or those done in pursuit of leisure. Let these turbulent times of information superhighways ignite excitement and candor in the human spirit with full text electronic information delivery as the catalyst!

Glossary of Terms

ABI/INFORM. A U.S. database containing business information.

Adapter Board. A printed circuit board that connects a computer system to peripherals or to devices that provide special functions.

Amplifier. A device which, by enabling a received wave of signals to control a local source of power, is capable of delivering an enhanced reproduction of the essential characteristics of the original wave.

Analog Transmission. Transmission of a continuously variable signal, as opposed to a discrete signal.

ANSI. Acronym for American National Standards Institute. An organization that sets standards for developing services and manufacturing products.

Archie. A database and related programs on the Internet which gives the user information about the contents of various archives. It facilitates location of files that are publicly available via anonymous FTP.

Artificial Intelligence (AI). Devices and applications that exhibit human intelligence and behavior including robots, expert systems, voice recognition, and natural and foreign language processing.

Asynchronous Transmission. Data transmission in which the transmission of a character or a block can begin at any time, but in which the bits that represent the character or block have equal time duration.

AT&T. American Telephone & Telegraph—The largest long distance communications carrier in the U.S.

Attenuation Rate. A rate at which a telecommunications signal loses intensity in the course of transmission.

Bandwidth. In data communications, the difference between the highest and lowest frequencies of a range of frequencies. It is expressed in Hertz and used to measure the capacity of a communications channel.

Baud Rate. In remote communications, the transmission rate that is synonymous with signal events. It is usually expressed in bits per second.

Bi-lateral. Pertaining to international agreements whereby two countries are involved as signatories.

BOC. Acronym for Bell Operating Company. One of the 22 telephone companies that was formerly part of AT&T and now one of the seven regional Bell telephone companies.

bps. Bits per second. In serial transmission, the instantaneous bit speed at which a device or channel transmits a character.

Bridge. A device that connects networks of the same type, allowing equipment on one local network (LAN) to communicate with devices on another.

BRS. Acronym for Bibliographic Retrieval Service, an online information service.

BSI. Acronym for British Standards Institute. An organization for setting standards for developing services or manufacturing products.

Byte. A string that consists of a number of bits, treated as a unit and representing a character.

CAD. Acronym for Computer Aided Design. Using computers to design products. CAD systems are high-speed workstations or personal computers using CAD software and input devices, such as graphic tablets and scanners.

CAE. Acronym for Computer Aided Engineering. Software that analyzes designs which have been created in the computer, or done elsewhere and then entered in the computer. Different types of analyses may be done such as structural analysis and electronic circuit analysis.

CAM. Acronym for Computer Aided Manufacturing. Automated manufacturing systems and techniques, that include numerical control, robotics, and material requirements planning (MRP).

Central Office. A telephone switching facility that interconnects subscribers' telephone lines to each other and to intra and intercity trunk lines.

Channel. In telecommunications or data communications, a path along which signals can be sent. Examples include output, input, and data and signaling channels.

Client. A functional unit on a network that receives shared services from a server.

Client/Server. In TCP/IP, the mode of interaction in distributed data processing in which a program at one site sends a request to a program at another site and awaits response. The requesting program is called the client. The answering program is called the server.

Coaxial Cable. A cable consisting of one conductor, usually a small copper tube or wire within and insulated from another conductor of a larger diameter. Usually encased in copper tubing or braided.

Connectivity. The capability of a system or device to be attached to other systems or devices without modifications to its basic operations.

COM Catalog. A catalog on microfilm, that has been created by data recorded directly from computer generated signals.

COMSAT. Acronym for Communications Satellite Corporation. An organization created by authorization of congress as part of the Communications Act of 1962. It is a private corporation (not an agency of the U.S. Government, although subject to government regulation) which was created primarily to establish and manage a commercial communications satellite system. COMSAT acts as manager of INTELSAT and also represents the U.S. on INTELSAT.

CWIS. Acronym for Campus-Wide Information System. A system that provides a college or university community with online information about class schedules, and special events, as well as providing directory information, calendars, bulletin boards, and databases.

Cyberspace. A world in which computers, telecommunications, and people coexist. The term was coined by William Gibson in his science fiction novel *Necromancer.* You are in cyberspace when you are connected online to an information superhighway—Internet, Bitnet and the like.

Datagram. In packet switching, a self-contained packet independent of other packets, that carries information sufficient for routing from the original data terminal equipment (DTE) to the destination DTE without relying on earlier exchanges between the DTEs and the network.

Data Terminal Equipment (DTE). A terminal or computer attached to a data network as an end-user node. Such a device must operate according to the defined protocol for the network.

Demodulate. To return a modulated signal to its original state.

Desktop Computer. A computer that will fit on the top of a regular office desk. Most personal computers are desktops.

DIANE. Acronym for Direct Access Network for Europe. It refers to information services offered over the Euronet system.

Dielectric. In Electrical Engineering, an insulating material or device that does not conduct electricity.

Digital Transmission. A mode of transmission in which all information is transmitted as serial discrete pulses.

Distributed Processing. Data processing in which some or all processing, storage, and control functions, in addition to input/output functions, are dispersed among data processing stations.

Downlink. Pertaining to data transmission from the head-end to a data station in satellite communications. The head-end is the satellite and the data station is the earth station.

Download. The process of transferring or receiving programs or data from one computer system to another.

EC. Acronym for European Community.

EIA. Electrical Industries Association. A membership organization, founded in 1924, that includes manufacturers of electrical and electronic equipment parts and systems.

EISA. (Extended ISA) PC bus standard that extends the AT bus (ISA bus) to 32 bits and provides bus mastering. It enhances the processing power of a PC-based CPU.

Electronic Bulletin Board. A computer system that maintains a list of messages so that people can call up with their computer and either post a message or read those already there.

Electronic Mail (e-mail). Correspondence in the form of messages transmitted among computer users over a computer network.

Euronet. A European packet-switching network for the transmission of digital information. It was established by the European Commission and has access points (nodes) in each of the member states.

Expert System. An interactive computer program that helps the users with problems that would otherwise need the assistance of human experts. It presents the computer as an expert on some topic.

Fiber Optics. A brand of optical technology concerned with the transmission of signals through fibers made of transparent materials such as glass, fused silicon, and plastic.

Flat File. A file that has no hierarchical structure. A list or table of items.

Freenet. Community-based bulletin board system supported by the National Public Telecommunications Network (NPTN). Similar to small towns on the electronic superhighway, freenets provide e-mail, discussion groups, local information, and sometimes connection to the Internet. One may visit a freenet, but in some cases a registration fee may be charged.

Front-end. A computer that handles the communications processing in a mainframe environment. In databank access, an interface that provides preliminary search connection protocols to facilitate ease of searching.

FTP. File Transfer Protocol, is a phrase used to describe the method of transferring files from one computer or terminal to another.

Gateway. A hardware/software combination that links systems with different protocols or standards. It is used to facilitate communication among networks.

Geostationary Orbit. The path occupied by a satellite that always remains constant with respect to the earth's surface, although the satellite and the earth are both rotating.

GIGO. Acronym for garbage-in-garbage-out. The concept means that products generated from unreliable data are equally unreliable.

GNP. Acronym for Gross National Product. An estimate of the total production of goods and services of a given country or nation.

Gopher. Client/server software which provides access to items on a network. It is menu-driven and used for exploring Internet resources.

GUI. Graphical user interface. Access to the computer that incorporates icons, pulldown menus, and a mouse. Examples include Macintosh, Windows, Presentation Manager (OS/2).

Hard-wired. Pertaining to the physical connection of two pieces of electronic equipment by means of cable.

Hertz. A unit of frequency equal to one cycle per second.

Histogram. A vertical bar chart often used to graph statistical information. The height of the bars are in proportion to the values of the data items they represent.

Host Computer. In a computer network, a computer that usually performs network control functions and provides end-users with services such as computation and database access.

HTTP. Acronym for Hypertext Transfer Protocol. A protocol used to transmit hypertext-based documents. Links among documents on the system allow users to refer to cited documents.

Icon. A graphic symbol, displayed on a screen, that a user can point to with a device such as a mouse in order to select a particular function or software application.

INTELSAT. Acronym for International Telecommunications Satellite Consortium. An international organization formed in 1964 to facilitate communication among several countries.

Intermediation. Pertaining to information retrieval systems, whereby a human intermediary searches the databank and provides information to an end-user.

Internet. A global network connecting thousands of disparate networks all over the world. It uses the TCP/IP protocol as the standard for transmitting information.

Interoperability. The capacity to communicate, execute programs, or transfer data among various units in a way that requires the user to have little or no knowledge of the unique features of those units.

ISA. Acronym for Industry Standard Architecture.

ISO. Acronym for International Organization for Standardization. An organization of national standards bodies established to promote development of standards to facilitate international exchange of goods and services.

Kilo. One thousand.

LAN. Acronym for Local Area Network. A communications network connecting devices which exchange signals in a small area. It normally has a server(s) and client receiver terminals.

Laptop Computer. A portable computer small enough to be operated on the desktop or in one's lap.

LED. Acronym for Light Emitting Diode, a commonly used alphanumeric display unit that glows when supplied with a specific voltage.

LISTSERV. An automated mailing list distribution system that started on the BITNET/EARN network. Listservs are now part of the Internet system. Users can post messages and get other peoples messages electronically mailed to them.

Logic. The systematized interconnection of digital switching functions, circuits, or devices.

MAN. Acronym for Metropolitan Area Network. A communications network covering a limited area.

Microwave. Any electromagnetic wave in the radio frequency spectrum above 890 megahertz. It is used in data communications.

MNC. Acronym for multi-national corporation. A corporation with marketing and product development interests geared to many countries

Modem. (Modulator/demodulator). A device which modulates and demodulates signals. One of the functions of a modem is to enable digital data to be transmitted over analog transmission facilities.

Modularity. The concept of designing computer systems in a building-block format to promote efficient and economical upgrading of equipment.

Modulate. To vary the amplitude, frequency, or phase of a signal.

Monitor. Synonym for video display terminal.

Multi-lateral. Pertaining to international agreements whereby more than one country is a participant or signatory.

Multimedia Device. A device capable of processing mixed media text graphics and sound originating from multimedia, such as compact disks, video disks, tapes, and cassettes.

Multiplexor. A device that allows several communications lines to share one data communications channel. Abbreviated as MUX.

MUX. Abbreviation for multiplexor.

Network. A configuration of data processing devices and software connected for information exchange.

Network Interface Card. An expansion card that connects a computer to a network.

Node. Any terminal, computer, or peripheral in a computer network that has legitimate access to the network.

NSFNET. One of Internet's primary networks. It connects sixteen nodes including five supercomputing sites in academic and research organizations across the U.S. The main trunk operates at 45 Mbps and connects several regional networks to the Internet.

OCLC. Online Computer Library Center. An organization that provides services to libraries and information centers, such as cataloging by copy, interlibrary loan records verification, and request forwarding.

Online Information. Information stored in a computer system that can be displayed, used, and modified in an interactive manner without the need for a hard copy.

Optical Fiber. Any filament made of dielectric material that guides light regardless of its ability to send signals.

OSI. Open Systems Interconnection. A networking standard for interconnecting disparate computer systems.

Packet. In data communications, a sequence of binary digits, including data and control signals, transmitted and switched as a unit.

Packet switching. A technique for handling high-volume traffic on a network by breaking messages into fixed-length packets that are transmitted to destinations via the most efficient route. Packets from the same original message may be sent by different routes.

Paging. The process of keeping program segments on a disk and calling them into memory as needed. The technique allows the computer system to move from real (internal) storage to virtual (auxiliary) storage.

Parallel Interface. A data transmission protocol, whereby all bits of data in a given byte (or word) are transferred simultaneously, using a separate data line for each bit.

Parallel Transmission. The simultaneous transmission of the group (of bits) representing a character or other entity of data.

PBX. Private Branch Exchange. An in-house switching system that interconnects telephone extensions in an organization to each other, as well as to outside telephone networks.

Peripheral. Pertaining to an operation or a device such as a tape, disk, or printer that is used to support the primary operations of the computer system.

Personal Computer. The smallest and least expensive class of computer. A computer designed to be used by one person at a time.

Portability. The capability of a program to be executed on various types of data processing systems without converting it to a different language and with little or no modification.

Post Processor. Software that provides some final processing to data. For example, formatting it for display or printing.

Protocol. A set of rules or conventions, implemented mainly by software, which allow programs on separate machines or networks to communicate.

RAM. Acronym for random access memory. It is the part of a computer used for temporary storage of data as it is being processed.

Realtime. Pertaining to online computer processing systems which receive and process data fast enough to produce output to control, direct, or affect the outcome of an ongoing activity or process. Airline ticket reservations are a good example.

Refresh Rate. In computer graphics, the rate at which a display image is refreshed in order to make it stable on a screen.

Relational Database. A database in which some data items in one type of record refer to records of a different type. Relational databases give the user the flexibility to link (join, or create a relationship between) information stored in many disk files. It allows the user to interchange and cross reference information between two different types of records. For example, comparing information in a group of invoices to information in an inventory file.

Repeater. In telecommunications, a device that amplifies or regenerates the data signal in order to extend the distance of transmission.

Router. A computer that determines the path of a network traffic flow. The path selection is made from several alternative paths, based on information obtained from specific protocols and algorithms which attempt to identify the shortest or best path, and other criteria, such as matrics and protocol specific destination addresses.

Satellite. Earth-orbiting device capable of relaying communications signals over long distances. A radio receiving/transmitting station in geosynchronous orbit 25,000 miles (35,965 kilometers) above the earth is used in satellite-based communications.

Search Engine. A set of programs used to search a comprehensive databank.

Serial Interface. A data transmission scheme that sends data and control bits sequentially over a single transmission line. With reference to a computer's serial input/output connection, the phrase usually implies the use of an RS-232 or RS-422 interface standard.

Serial Transmission. The sequential transmission of the signal elements of a group (bits) representing a character or other entity of data.

Server. A functional unit that provides sharable services to workstations over a network, for instance a file server, a print server, or a mail server.

SMTP. Simple Mail Transfer Protocol is a method of processing mail. Used on the Internet under the TCP/IP protocol suite.

SNA. Acronym for System Network Architecture, a proprietary IBM suite of protocols which allows communication among IBM devices.

Sound Card. In multimedia, an add-on adapter board that incorporates a synthesizer without a musical keyboard and has audio output jacks for the sound created.

Standard. Guide used to establish uniform practice and common techniques.

Sync Signals. A signal transmitted to establish character synchronization in synchronous transmissions.

Synchronization. Adjustment of the chronological relationships between events, either to cause them to coincide or to maintain a fixed-time difference.

Synchronous Transmission. In telecommunications, data transmission in which the bits are transmitted at a fixed rate. Both the transmitter and receiver use the same clock signals for synchronization.

T1. The term T1 refers to the transmission lines along which digital signals travel at 1.544 Mbps, equivalent to 48 voice channels.

T2. A common carrier's facility that transmits digital signals at 6.312 Mbps, equivalent to 96 voice channels.

T3. A common carrier's facility that transmits digital signals at 44 Mbps, equivalent to 672 voice channels.

T4. A common carrier's facility that transmits digital signals at 273 Mbps, equivalent to 4,032 voice channels.

TCP/IP. Transmission Control Protocol/Internet Protocol. A set of communications protocols that support peer-to-peer connectivity functions for both local and wide area networks.

Telecommunications. Transfer of data or information from one place to another over long distance, using telephone lines, microwave, and/or satellites.

Telecommuting. Working at home with telecommunications links to the office.

Teleconferencing. A conference among people remote from each other linked by telecommunications devices. An alternative to travel and face-to-face meetings. A teleconference may be conducted with two-way video, audio, data, and facsimile transmission.

Time-sharing. A method of using a computer system that allows a number of users to execute programs concurrently and to interact with the programs during execution.

Topology. Physical layout of a computer network. Interconnection of devices and communications channels into a network configuration.

Transborder Data Flow (TBDF). Pertaining to data communications between computer systems located across national borders.

Transceiver. Any device (terminal) that can transmit and receive signals or data.

Transmission Medium. The physical medium that conveys signals between data stations. Examples include twisted pair wire, optical fiber, and coaxial cable.

Transparent. Pertaining to any process that is not visible to the user. For instance, how a file is stored on tracks and sectors on a disk are transparent to the user.

UNIX Operating System. An operating system developed by Bell Laboratories that features multi-programming in a multi-user environment. UNIX was originally developed for minicomputers, but has been adapted for mainframes and microcomputers.

Uplink. Pertaining to data transmission from a data station to the head-end, which is the orbiting satellite.

Upload. To transfer or send programs or data from one computer to another.

User Friendly. A system with which relatively untrained users can interact easily.

User Interface. Hardware, software, or both that allows a user to interact with and perform operations on a system, program, or device.

Veronica. Acronym for Very Easy Rodent-Oriented Net-Wide Index to Computerized Archives. An Internet utility for searching Gopher sites by keyword.

Video RAM. A special type of dynamic RAM used in high-speed video applications.

Virtual Terminal. A generalized logical model of different terminals of a certain class, describing how terminals of that class will perform in an open systems interconnect (OSI) environment.

Virus. In computer security, a self-perpetuating program which infects and may damage another program or set of programs.

WAN. Acronym for Wide Area network. A communications network that cover vast areas.

Web Browser. A software package that enables users to search and manipulate information on the World Wide Web. Examples are MOSAIC and Netscape.

Workstation. A microcomputer connected to a network or standalone with local processing capacity. It allows the user to access information resources needed to perform most activities associated with his/her job. Activities may include: accessing remote databases, statistical analysis, generating reports, and sending or receiving electronic mail.

Selected Bibliography

Adeyemi, Nat M. "Issues in the Provision of Information to Developing Countries." *African Journal of Library, Archive and Information Science* 1 (April 1991): 1–8.

Aidarous, Salah and Thomas Plevyak eds. *Telecommunications Network Management Into the 21st Century.* New York: IEEE Press, 1994.

Albiges, Luce-Marie. "Remote Public Access to Picture Databanks." *Ifla Journal* 18 (1992): 37–42.

Ardoin, Beth and William Weems. "Campus-Wide Information Systems: User Publication of Documents." *Cause/Effect* (Winter 1993): 21–26.

Baker, Richard H. *Network Security: How to Plan for It and Achieve It.* New York: McGraw-Hill, 1995.

Bar, Fernando L. "Can Third World Countries Become Information Societies?" *International Journal of Micrographics and Optical Technology* 10 (1992): 71–73.

Basch, Reva. *Electronic Information Delivery: Ensuring Quality and Value.* Aldershot, UK: Gower, 1995.

Bellamy, John. *Digital Telephony, 2nd ed.* New York: John Wiley, 1991.

Bender, David R. "A Strategy for International Information Policy." *Libri* 43 no. 3 (1993): 210–231.

Bennett, Valerie. "Electronic Document Delivery Using the Internet." *Bulletin of Medical Library Association* 82 no. 2 (April 1994): 163–167.

Bernbom, Gerald. "Campus-Wide Information Systems: Managing Information Content." *Cause/Effect* (Winter 1993): 3–5.

Bing, Jon. "Reflections on EC Data Policy." *Transnational Data and Communications Report* (May 1990): 15–17.

Black, Uyless. *Computer Networks, Protocols, Standards, And Interfaces.* Englewood Cliffs, N.J.: Prentice-Hall, 1987.

———. *Data Link Protocols.* Englewood Cliffs, N.J.: Prentice-Hall, 1993.

———. *Data Networks: Concepts, Theory and Practice.* Englewood Cliffs, NJ: Prentice-Hall, 1989.

Bluh, Pamela. "Striking a Balance: Document Delivery in the Nineties." *Law Library Journal* 85 (1993): 599–608.

Bollinger, William A. "Information Access Policies in the 1990s: National and International Concerns." In Martha Williams ed. *Proceedings of the Eleventh National Online Meeting.* Medford, NJ: Learned Information, Inc., 1990.

Brentrup, Robert. J. "Building a Campus Information Culture." *Cause/Effect* (Winter 1993): 8–14.

Brockman, John R. "Information Management and Corporate Total Quality." *Journal of Information Science* 19 (1992): 259–266.

Brown, J.S. "Research That Reinvents the Corporation." *Harvard Business Review* 108 (January/February, 1991): 330–339.

Buck, R.G. "ISDN and IN Working Together." *BT Technology Journal* 13, no. 4. (April 1995): 113–119.

Buga, Wladyslaw J. "AT & T International Services Support Architecture Overview." In *IEEE 1992 Network Operations and Management Symposium.* New York: IEEE, 1992.

Burger, Robert. *Information Policy: A Framework for Evaluation and Policy Research.* Norwood, NJ.: Ablex Publishing Corp, 1993.

Burton, Hilary D. "The Livermore Intelligent Gateway: An Integrated Information Environment." *Information Processing and Management* 25 no. 5 (1989): 509–514.

Callendar, Mike. "Future Public Land Mobile Telecommunication Systems." *Canadian Journal of Electrical and Computing Engineering* 17 no. 4 (1992): 152–155.

Campbell, Robert. "Document Delivery and the Journal Publisher." *Scholarly Publishing* (July 1992): 213–221.

Chartrand, R. L. and J. Bortnick. *Information Policy and Technology Issues: Public Laws of the 95th Through 101st Congresses, CRS Report 91–569 SPR, Congressional Research service.* Washington, D.C.: Library of Congress, 1991.

Churbuck, David C. and Jeffrey S. Young. "The Virtual Workplace." *Forbes* 150 no. 12 (1992): 184–190.

Clark, Martin P. *Networks and Telecommunications: Design and Operation.* New York: John Wiley, 1991.

Cleveland, Gary. *Research Networks and Libraries: Applications and Issues for a Global Information Network.* London: Ifla, 1991.

Committee on Applications and Technology. *What It Takes to Make It Happen: Key Issues For Application of The National Information Infrastructure.* Washington, D.C.: Information Infrastructure Task Force, 1994.

Corey, James F. "A Grant for Z39.50," *Library Hi Tech* 12 no. 1 (1994): 37–47.

Cox, Donald C. "Wireless Network Access for Personal Communications." *IEEE Communications* (December 1992): 96–97.

Derfler, Frank J. Jr. *PC Magazine Guide to Connectivity, Second Edition.* Emeryville, CA: Ziff-Davis Press, 1992.

Doctor, Ronald D. "Seeking Equity in the National Information Infrastructure." *Internet Research* 4 no. 3 (1994): 9–22.

Efthimiadis, Efthimis N. "Online Searching Aids: A Review of Front Ends, Gateways and Other Interfaces." *Journal of Documentation* 46 no. 3 (1990): 218–262.

Elias, A.W. "Copyright, Licensing Agreements and Gateways." *Online Review* 14 no. 4 (1990): 225.

Evans, Ed. "Enabling South Africa: Developing of an Intelligent Gateway." *The Electronic Library* 11 no. 6 (1993): 279–282.

Figueredo, Nice. "Information as a Tool for Development." *The International Information and Library Review* 24 (1992): 189–201.

Finnegan, Gregory. "Wiring Information to a College Campus: A Port for Every Pillow." *Online* 14 no. 2 (March 1990): 37–40.

Fleming, Maureen. "Microsoft Network and Interchange: Changing Online Metaphors." *Online* 19 no. 2 (1995): 71–72.

Frankel, David. "ISDN Reaches the Market." *IEEE Spectrum* 32 no. 6 (June 1995): 20–25.

Furey, Tim R. et. al. "Applying Information Technology to Reengineering." *Planning Review* 21 no. 6 (November/December 1993): 22–25, 55.

Gale Directory of Databases, Volume 1: Online Databases July 1994, Edited by Kathleen Young Marcaccio. Washington, D.C.: Gale Research Inc., 1994.

Galvin, Thomas J. "Leadership in Legislation and Public Policy Development: The Case of the American Library Association." *Library Trends* 40 no. 3 (1992): 431–56.

Garrett, J.R. "Digital Libraries: The Grand Challenges." *Educom Review* 28 no. 4 (July/August 1993): 17–21.

Gassaway, Laura N. "Document Delivery." *Computers in Libraries* 14 no. 5 (May 1994): 25–32.

Glory, Blaise. "Managing Information Resources in a Telecommuting Environment." *Special Libraries* 85 no. 1 (1994): 30–34.

Gosztony, Geza. "CCITT Work in Teletraffic Engineering." *IEEE Journal on Selected Areas in Communications* 9 no. 2 (1991): 131–134.

Hallgren, Martyne M. "Funding an Internet Public Good: Definition and Example." *Computer Networks and ISDN systems* 27 no. 3 (1994): 403–409.

Hammer, Michael. "Reengineering Work: Don't Automate, Obliterate." *Harvard Business Review.* (July/August 1990): 104–113.

Hammer, Michael and James Champy. *Reengineering the Corporation: A Manifesto for Business Revolution.* New York: Harper Bush, 1992.

Heldman, Robert K. *Information Telecommunications: Networks, Products & Services.* New York: McGraw-Hill, 1994.

Hernon, Peter. "Equity in Public Access to Government Information." *Government Information Quarterly* 10 no. 3 (1993): 301–304.

———. "Government Information Principles." *Government Information Quarterly* 8 no. 4 (1991): 393–399.

Hernon, Peter and Charles McLure. *Public Access to Government Information: Issues Trends and Strategies.* Norwood, NJ: Ablex Publishing Corp, 1984.

Holm, Liv A. "Connectivity and Protocols—The Technical Side: OSI and TCP/IP, FTP, TELNET, SR, ILL, Update." *Ifla Journal* 20, no. 2 (1994): 158–170.

Howe, Eric. "The United Kingdom's Data Protection Act." *Government Information Quarterly* 8 no. 4 (1991): 345–357.

Hudson, Heather E. *Communication Satellites: Their Development and Impact.* New York: The Free Press, 1990.

Hunter, John D. and William W. Ellington. "ISDN: A Customer Perspective." *IEEE Communications Magazine* (August 1992): 12–23.

Jackson, A.E. "A Retail Opportunity in the Contemporary Information Market." *Aslib Proceedings* 44, no. 10 (October 1992): 341–349.

Jackson, Mary E. "Integrating Document Delivery Services With Electronic Document Delivery Technologies." *Law Library Journal* 85 (1993): 609–618.

Johnson, Jean M. "Asa on the Bell Atlantic Gateway: The Lehigh Experience." In Martha E. Williams ed. *Proceedings of the National Online Meeting.* Medford, N.J.: Learned Information, Inc., 1990.

Kehoe, Brendan P. *Zen and the Art of the Internet.* Englewood Cliffs, NJ: Prentice-Hall, 1994.

Kelly, Brian. "Becoming an Information Provider on the World Wide Web." *Computer Networks and ISDN Systems.* 27 no. 3 (1994): 353–360.

Kessler, Gary C. *ISDN: Concepts, Facilities, and Services, 2nd Ed.* New York: McGraw-Hill, 1993.

Keyes, Jessica. *Infotrends: The Competitive Use of Information.* New York: McGraw-Hill, 1993.

Khalil, Mounir. "Document Delivery: A Better Option?" *Library Journal* 118 no. 2 (February 1, 1993): 43–47.

Kibirige, Harry M. "Electronic Information Delivery: A Study of the U.S. Infrastructure." In Martha Williams ed. *National Online Meeting.* Medford, NJ: Learned Information, Inc., 1994.

———. "Information Communications Highways in the 1990s: An Analysis of Their Potential Impact on Library Automation." *Information Technology and Libraries* 10 no. 3 (September 1991): 172–184.

Kinnucan, Mark T. "Demand for Document Delivery and Interlibrary Loan in Academic Settings." *Library and Information Science Research* 15 (1993): 355–374.

Kluepfel, Henry M. "Securing A Global Village and Its Resources." *IEEE Communications Magazine* (September 1994): 82–89.

Kovacevic, Antonio and Nicolas Majluf. "Six Stages of IT Strategic Management." *Sloan Management Review* 34 no. 4 (1993): 77–87.

Kraemer, Kenneth L. et al. "Economic Development, Government Policy, and the Diffusion of Computing in Asian-Pacific Countries." *Public Administration Review* 52 no. 2 (March 1992): 146–156.

Kriegel, Robert. *If It Ain't Broke, Break It.* New York: Warner Books, 1991.

Langenberg, Donald N. "Information Technology and the University: Integrated Strategies for the Twenty-First Century." *Journal of the American Society for Information Science* 45, no. 6 (1994): 323–325.

Leach, Ronald G. and Judith E. Tribe. "Electronic Document Delivery: New Options for Libraries." *Journal of Academic Librarianship* 18 no. 6 (1993): 359–364.

Learn, Larry L. "The Role of Telecommunications in Library Automation: Past, Present and Future." *Library Technology Reports* (July-August 1990): 503–515.

Lee, Sunro and Richard Leifer. "A Framework for Linking the Structure of Information Systems With Organizational Requirements for Information Sharing." *Journal of Management Information Systems* 8 no. 4 (Spring 1992): 27–44.

Leon, R. E. "The Internet World: Its Protocols and Mainstream Services." *Online & CD-ROM Review* 18 no. 4 (1994): 229–239.

Lievrouw, Leah A. "Information Resources and Democracy: Understanding the Paradox." *Journal of the American Society for Information Science* 45, no. 6 (1994): 350–356.

Lloyd, Les. *Campus-Wide Information Systems and Networks.* Westport: Meckler, 1992.

Lynch, Clifford. "Networked Information: A Revolution in Progress." In, Brett Sutton and Charles H. Davis. *Networks, Open Access, and Virtual Libraries.* Urbana-Champaign: University of Illinois Press, 1992.

MacBride, S. *Many Voices, One World: Towards a New More Just and More Efficient World Information and Communication Order.* New York: Unipub, 1980.

Mantviloff, Victor. *National Information Policies: A Handbook on the Formulation, Approval, Implementation and Operation of a National Policy on Information.* Paris: UNESCO, 1990.

Marchand, Donald A. "Infotrends: A 1990s Outlook on Strategic Information Management." *Information Management Review* 5, no. 4 (1991): 23–32.

Martinez, Michael E. "Computers and Social Equity: Views From the National Assessment of Educational Progress (NAEP)." In *Proceedings of The 55th ASIS Annual Meeting of The American Society for Information Science, Pittsburgh, PA. October 26–29.* Debrah Shaw ed. Medford, NJ: Learned Information, Inc., 1992.

Masmoudi, M. "The New Information Order." *Journal of Communication* 29 (1979): 172–185.

McGovern, Tom. *Data Communications: Concepts and Applications.* Scarborough, Ontario, Canada: Prentice-Hall, 1988.

McMullen, Melanie ed. *Networks 2000: Internet, Information Superhighway, Multimedia Networks and Beyond.* San Francisco: Miller Freeman Books, 1994.

McMundo, George and Evelyn Simpson. "The White House Files." *Journal of Information Science* 20 no. 5 (1994): 305–313.

Miflin, Ingrid and Jean Williams. "Online Catalog Maintenance: The Role of Networks, Computers, and Local Institutions." *Information Technology and Libraries* 10 no. 4 (1991): 363–274.

Mitchell, Eleanor and Sheila A. Walters. *Document Delivery Services: Issues and Answers.* Medford, NJ: Learned Information, Inc., 1995.

Mokhtarian, Patricia L. and Koji Sato. "A Comparison of the Policy, Social, and Cultural Contexts for Telecommuting in Japan and the United States." *Social Science Computer Review* 12 no. 4 (1994): 641–658.

Morris, Ruth C.T. "Toward a User-Centered Information Service." *Journal of the American Society for Information Science* 45 no. 1 (1994): 20–30.

Morton, Michael Scott. *The Corporation of the 1990s: Information Technology and Organizational Transformation.* New York: Oxford University Press, 1991.

Network Protocols, International Conference on: Proceedings. San Francisco, CA: IEEE Computer Society Press, 1993.

Newberg, Paula R. *New Directions in Telecommunications Policy.* Durham, NC: Duke University Press, 1989.

Nugter, A.C.M. *Transborder Data Flow of Personal Data Within the EC.* Boston (and Deventer—The Netherlands), Kluwer Law and Taxation Publishers, 1990.

Nulty, Peter. "When to Murder Your Mainframe." *Fortune* 128 no. 11 (November 1, 1993): 109–120.

Offer, Colin. "Charging for Information: A Changing Climate and an Emerging Market Place." *Informediary* 5 no. 1 (1991): 49–62.

Olson, Mark. "We Must Apply the Principles of Business Process Reengineering in Higher Education." *Cause/Effect* (Winter 1993): 54.

Organization for Economic Co-operation and Development. *Information Networks and New Technologies: Opportunities and Policy Implications for the 1990s. Information Computer Communications Policy No 3.* Paris: OECD, 1992.

Perez, Victor L. "Developing Informatics Policies." *Information and Management* 23 (1992): 7–14.

Peters, Paul Evan. "Networked Information and Services: Next Steps on the Road to the Distributed Digital Libraries of the Twenty-first Century." In Bret Sutton and Charles H. Davis. *Networks, Open Access, and Virtual Libraries: Implications for the Research Library.* Urbana-Champaign: University of Illinois Press, 1992.

Pinto, Jeffrey K. *Successful Information System Implementation: The Human Side.* Upper Darby, PA: Project Management Institute, 1994.

Plesser, Ronald L. and Emilio W. Cividanes. "EC Personal Data Privacy: US Concerns." *Transnational Data and Communications Report* (November 1990): 19–20.

Porter, James H. "Business Reengineering in Higher Education: Promise and Reality." *Cause/Effect* (Winter 1993): 48–53.

Pugesley, William and Rickye Lennon. "A Report on the Status of Fee-Based Information Brokering in the United States." *Informediary* 4 no. 1 (1990): 13–32.

Pullinger, David J. "Learning From Putting Electronic Journals on SuperJANET: The SuperJournal Project." *Interlending & Document Supply* 23 no.1 (1994): 20–27.

Romei, Laura K. "Telecommuting: A Workstyle Revolution?" *Modern Office Technology* 37 no. 5 (1991): 38–40.

Sauvant, Karl P. *International Transactions in Services: The Politics of Transborder Data Flows.* Boulder, Co: Westview Press, 1986.

Schoenbrun, Cynthia. "Easynet: What Has Become of the Small Giant?" *Online* (January 1993): 52–56.

Schoonmaker, Sara. "Trading Online: Information Flows in Advanced Capitalism." *The Information Society* (1993): 39–49.

Sharpe, Diana M. "Asia-Pacific TDF Issues." *Transnational Data and Communication* (May 1990): 220–24.

Sheldon, Tom ed. *Lan Times Guide to Interoperability*. New York: Osborne McGraw-Hill, 1994.

Sprehe, J.T. *The Year in Federal Information Policy: Events and Commentary:14–15 (IDC Government Report)*, Washington, D.C.: Government Printing Office, 1993.

Srinavasan, S. "A Comparative Study of National ISDN Versus International ISDN Installations and Tariffs." *IEEE Network* 9 no. 3 (May/June 1995): 22–26.

Stallings, William. *Computer Communications: Architectures, Protocols, and Standards*. Los Alamitos, CA: IEEE Computer Society Press, 1992.

Summerhill, Craig A. "Connectivity and Navigation: An Overview of The Inter-Networked Information Infrastructure." *Ifla Journal* 20 no. 2 (1994): 147–157.

Swain, Leigh and Paula Tallim. "The Interlibrary Loan (ILL) protocol: Progress and Project." *IFLA Journal* 8 no. 4 (1992): 324–332.

Sylvia, Margaret. "Building a Gateway for the CD-ROM Network: A Step Toward the Virtual Library With the Virtual Microsystems V-Server." *Microcomputers for Information Management* 10 no. 2 (June 1993): 119–129.

Tolly, Kevin. *IBM Lan Gateways*. New York: McGraw-Hill, 1993.

Tonn, Bruce E. "Using the National Information Infrastructure for Social Science, Education, and Informed Decision Making." *Social Science Computer Review* 12 no. 2 (1994): 166–182.

Tropiano, Lenny and Dinah McNut. "How to Implement ISDN." *Byte* 20, no. 20 (April 1995): 67–68, 70, 72, 74.

Turkheimer, Frank. "Privacy and the Internet: The Next Step." *Computer Networks and ISDN Systems* 27 no. 3 (1994): 395–401.

U.S. Congress. Committee on Science, Space, and Technology. Subcommittee on Technology, Environment, and Aviation. 103 Congress, Second Session. *Electronic Commerce and Interoperability in the National Information Infrastructure.* Washington, D.C.: Government Printing Office, 1994.

U.S. Congress, Office of Technology Assessment. *Making Government Work: Electronic Delivery of Federal Services.* Washington, D.C.: Government Printing Office, 1993.

U.S. Congress, Office of Technology Assessment. *U.S. Telecommunications Services in European Markets, OTA-TCT-548.* Washington D.C.: U.S. Government Printing Office, 1993.

U.S. Department of Commerce. *National Information Infrastructure: Agenda for Action.* Washington, D.C.: U.S. Department of Commerce, 1993.

Wallestein, Gerd. *Setting Global Telecommunications Standards.* Norwood, MA: Artech House, 1990.

Wang, P. "Information Systems Solutions For Transborder Data Flows for Multinational Companies." *International Journal of Information Management* 13 (1993): 29–40.

Welsch, Erwin K. and Ellen Shultz. "Developing Gateways to Online Information Utilities." *Library Software Review.* (Mar/Apr 1990): 83–90.

Wicks, Wendy ed. *Government Information and Policy: Changing Roles in a New Administration.* Philadelphia: The National Federation of Abstracting and Information Services, 1994.

Williams, Martha E. "The Internet: Implications for the Information Industry and Database Providers." *Online & CDROM Review* 18 no. 3 (1994): 149–156.

Willis, Katherine et al. "TULIP—The University Licensing Program: Experience at the University of Michigan." *Serials Review* (Fall 1994): 39–47.

Wilson, Thomas C. ed. *Impact of Technology on Resource Sharing: Experiment and Maturity.* New York: Haworth Press, 1992.

Young, J.A. "The Advantages of Telecommuting." *Management Review* 80 no. 7 (1991): 19–21.

Appendix A
Document Delivery Services*

ADONIS
238 Main Street
Cambridge, MA 02142
Tel: 800-944-6415

On CD-ROM, covers journal articles from 1991 to the present. Biomedical specialization. Contact: Chris Davis.

Bernan/Unipub
4611-F Assembly Drive
Lanham, MD 20706
Tel: 301-459-7666

Largest distributor of U.S. government-related publications. American agent for British HMSO publications. Also has UN, ILO, and World Bank documents. Order may be made by fax, e-mail, Internet, and telex.

The British Library (BL)
Document Supply Centre
Boston Spa, Wetherby, W. Yorkshire
L5237BQ, United Kingdom
Tel: 44-937-456060

Large supplier of documents, very broad subject coverage, humanities, social sciences, physical sciences, biological sciences. Contact: Customer Service.

Canada Institute for Scientific and Technical Information
Ottawa, Canada K1A OS2
Tel: 613-993-2350

Covers: business, science, technology, engineering, medicine, physical sciences, biological sciences. Contact: Kathryn J. Mikoski.

---

* The information in this Appendix was verified when the book went to press.

Chemical Abstract Service (CAS)
2540 Olentangy River Road
Columbus, OH 43210-0012
Tel: 614-447-3670

Covers: chemistry, chemical engineering. May be ordered through OCLC, Knight-Ridder, and STNORDER. Contact: Marian Nichol.

Disclosure, Inc.
5161 River Road
Bethesda, MD 20816
Tel: 800-633-8241

Covers: general business, investment, and law. May be ordered through Knight-Ridder, RLIN, e-mail. Contact: Kathy Cook or Monica Callahan.

Docutronics Information Services
130 W. 42nd St. 2nd Floor
New York, NY 10036
Tel: 800-238-3458

Covers general business, investment, newspapers. Contact: Kate Purcell.

ERIC Document Services, Inc.
Educational Resources Information Center
1301 Piccard Drive, Suite 300
Rockville, MD 20850-4305
800-799-3742

Covers: broad-based educational literature. May be accessed through OCLC, Dialog, and e-mail. Contact: Lisa Douds.

ESL Information Services
Engineering Societies Libraries (ESL)
345 E. 47th St.
New York, NY 10017
Tel: 212-705-7000

Covers: Engineering with emphasis on conference proceedings. May be accessed through OCLC, RLIN, Dialog. Contact: Customer Service.

Faxon Research Services
238 Main Street, Suite 201

Cambridge, MA 02142
Tel: 617-354-7112
Collaborative effort, Faxon Research Services and OCLC. Covers: business, science, technology, and medicine. Contact: Susan Sterns.

GeoREF Document Delivery Service
Americal Geological Association
4220 King Street
Alexandria, VA 22032
Tel: 703-379-2480
Provides photocopies of earth sciences documents available at U.S. Geological Survey Library in Reston, VA, the Library of Congress, and the American Geological Institute. Contact: Mary Ellen Shook.

Information for Business
2608 Ninth Street
Berkeley, CA 94710
Tel: 510-644-2111
Market research reports and competitive business intelligencce. Documents may be ordered by fax, e-mail, Internet. Delivery by fax, regular, and express mail. Contact: Larry Marks

Information Handling Services inc.
15 Inverness Way East
Englewood, CO 80150
Tel: 303-790-0600
World's largest technical and regulatory information provider. Information in electronic and microform formats—CD-ROM. Orders may be done by fax, e-mail, and Internet. Contact: Jean Wgner.

International Patent Research Office B.V.
P.O.Box 16260
Huygensstraat 23
2500 BG The Hague, Holland
The Netherlands
Tel: 31-70-388-9303
Performs patent searches and supplies copies of patents for the following countries: USA, Germany, France, Switzerland, Canada, Great Britain, Austria, and Sweden. Contact: Mr. S. Loeven.

ISI Genuine Articles
Institute for Scientific Information
3501 Market Street
Philadelphia, PA 19104
Tel: 800-336-4474
Covers academic journals—science, technology, medicine, humanities, and social sciences. May be accessed through Dialog, OCLC. Contact: Lisa Smith.

K R Source One
469 Union Ave.
Westbury , NY 11590
Tel: 800-238-3458
Collaborative effort between Knight-Ridder (formerly Dialog) and EI. Provides full text copies of engineering science documents. Contact: Lisa Hill.

Knight-Ridder Information, Inc.
El Camino Real
Mountain View, CA 94040-1400
Tel: 800-633-8241
Database vendor with full text databases in several subjects. Information searched and retrieved instantly. Contact: Diane Leo.

LEXIS Document Services
A Member of the Reed Elsevier plc group
Subsidiary of Lexis Nexis Online
801 Adlai Stevenson Drive
Springfield, IL 62703-4261
Tel: 800-227-9597
Legal and general information in full text. Provides online and offline delivery. Latter may be delivered by express or regular mail. Contact: Rebecca Cantrall.

Marketing Services Research Center
Electronic Industries Association
200 Pennsylvania Avenue NW
Washington, D.C. 20006-1813
Tel: 202-457-8739
Provides documents from its own sources—custom databases, books, and market research reports. Orders by e-mail, fax, and Internet. Contact: Michelle Bing.

MATH DOC
Mathematical Reviews
P.O. Box 8604
416 Fourth Street
Ann Arbor, MI 48107
Tel: 313-996-5268

A document delivery service to online users of the MATH/SCI, the database for the mathematical sciences. Covers many hard-to-find articles. May be ordered via Dialog and e-mail. Contact: Beverly Demchuk-Burke.

National Technical Information Service (NTIS)
5285 Port Royal Road
Springfield, VA 22161
Tel: 703-487-4650

A government agency under the Department of Commerce. Covers broad subject area of U.S. government sponsored research material. It also includes some foreign government supported research documents. Delivery may be by express mail or fax.

National Translation Center
Document Delivery
Library of Congress
3rd and A Sts. SE, Adams Building
Washington, D.C. 20540
Tel: 202-707-0100

A translation service on demand for unpublished translations in any language into English. Covers current journals, patents, conference proceedings, and standards. Delivery by express mail or fax.

NewsBank Readex
5020 N. Tamiami Trail, Suite 110
Naples, FL 33940
Tel: 800-762-8182

Covers material from newspapers, magazines with a broad subject coverage: business, education, humanities, social sciences, and medicine. Delivery may be by express mail or fax. Contact: Janet Marzulo

OCLC, Inc.
6565 Frantz Road

Dublin, OH 43017-3395
Tel: 800-848-5878

Index and contents databases provide the first steps in the search process. OCLC's World Catalog, Article First and Contents First give the initial citations. Full text documents are provided in collaboration with University Microfilm International (UMI) under its Article Clearinghouse. Contact: Customer Service.

Research Libraries Group
1200 Villa St.
Mountain View, CA 94041
Tel: 800-537-7546

Full text copies are available on demand. Delivery may be by fax, regular, or express mail. Ariel, an Internet-based delivery system provides electronic full text among organizations that support the Ariel software. Subject coverage is broad based. Contact: Kristin Taque.

UMI Info Store
Universal Microfilms International
500 Sansome St., Suite 400
San Francisco, CA 94111
Tel: 800-248-0360

Broad subject coverage, journals, magazines, items from dissertation abstracts, newspapers, and government documents. May be ordered via OCLC, Dialog and delivery is by regular or express mail or fax. Contact: Egils Kronlins.

UnCover 2
CARL Systems
3801 E. Florida, Suite 300
Denver, CO 80210
Tel: 303-758-3030

Covers several subjects from journal articles. Can be searched online and orders placed. Delivery may be by fax, express mail or Internet. Contact: Martha Whittaker.

Universal Serial and Book Exchange
2969 W. 25th St.
Cleveland, OH 44113
Tel: 216-241-6960

Provides full text from back issues of serial publications. May be ordered from OCLC and delivery by regular or express mail and fax. Contact: John T. Zubal.

Index

D